STUDY, MEASURE, EXPERIMENT

Allen King teaching an undergraduate optics laboratory, 1947

Study, Measure, Experiment

Stories of scientific instruments at Dartmouth College

David Pantalony

Richard L. Kremer

Francis J. Manasek

Terra Nova Press

Norwich, Vermont
Distributed by University Press of New England

ISBN 0-9649000-9-2

FIRST EDITION
First printing 2005

Dedicated to our teachers.

Acknowledgments

The authors are indebted to many people who have given generously of their time and knowledge.

We acknowledge particular debt to Allen King, who well after he stopped participating actively in the collection, continued to share his vast knowledge and to infect us all with his enthusiasm.

The Department of Physics and Astronomy provides a physical home for the collection. Bill Doyle and Gary Wegner willingly shared their knowledge and memories. Mary Hudson's and John Thorstensen's support was essential. The care that Tom Kenyon and Ralph Gibson have given the collection has, on many levels, left an indelible imprint. The undergraduate students in our course, Physics 7, who utilize instruments from this collection, gave us fresh insights and infused us with their boundless energy.

The riches of the Rauner Special Collections and the Dartmouth libraries proved invaluable for this project. We thank the directors of Special Collections, first Philip N. Cronenwett and then Jay Satterfield, and their extraordinary staff who unfailingly gave us their assistance. Without their help this book could not have been written.

Jere Daniell and Joseph Cullon of the History Department shared their broad knowledge of colonial America.

Otmar Foelsche, Director, Humanities Resources, assisted us with digital imaging.

Deborah Haynes, Kellen Haak and Katherine Hart (Hood Museum of Art) provided guidance, support and knowledge for cataloguing, conservation, and for the installation of student exhibits.

Deborah Jean Warner, Steve Turner, Peggy Kidwell, Roger Sherman and Harold Wallace (Smithsonian Institution, Washington, D.C.), Paolo Brenni (Istituto e Museo di Storia della Scienza, Firenze), Neil Brown (Science Museum, London), Randall Brooks (National Museum of Science and Technology, Ottawa), Robert Arns and David Hammond (University of Vermont), Julian Holland (Macleay Museum, Sydney), Don Osterbrock, (Lick Observatory, University of California), Sara Schechner (Harvard's Collection of Historical Scientific Instruments), Tom Greenslade (Kenyon College), Ellen Kuhfeld (Bakken Museum, Minneapolis), John Jenkins (American Museum of Radio and Electricity, Bellingham, Washington), Alison Morrison-Low (National Museums of Scotland, Edinburgh), Shae Trewin (Yale Historical Scientific Instruments), Karen Papineau (Currier Museum of Art, Manchester, New Hampshire), Elizabeth Cavicchi, Andrew Bell, John Briggs, Tom Suarez, Ronald Smeltzer, Ray Bates and Ray Giordano engaged us in fruitful discussions, clarifying many issues. We appreciate the generous sharing of their knowledge.

The Provost of Dartmouth College provided postdoctoral support that enabled one of us (DP) to spend eighteen months cataloguing the collection.

This book is based, in part, upon work supported by the National Science Foundation under grant number 0236341. Any opinions, findings, and conclusions expressed in this work are those of the authors and do not necessarily reflect the views of the National Science Foundation.

Table of Contents

INTRODUCTION

The Allen King Collection of Scientific Instruments contains about three thousand artifacts that were once employed for study, measure and experiment by faculty and students at Dartmouth College. When linked to related documentary materials--invoices and correspondence, trade catalogues, instruction manuals, inventories, laboratory notebooks, lecture notes, drawings and photographs--the instruments provide a rich cultural history of teaching and research in science at one of America's oldest educational institutions. Unlike collections assembled by private collectors or curators at smaller specialized or larger national museums, the King Collection preserves objects, the overwhelming majority of which are indigenous to Dartmouth. The changing composition of the collection over the past two centuries thus reflects not only local needs but also trends in the international trade of scientific instruments. And with its often idiosyncratic patterns of preservation and loss, the collection illustrates both the fragility and robustness of the material culture of science at an institution where adequate funds and storage space all too often were in short supply.

This book is not a comprehensive catalogue of the King Collection. Rather it presents about one hundred objects, selected to illustrate the diversity of artifacts in the collection and the range of stories evoked by these artifacts. We did not select only the most photogenic instruments, or only those with the most impressive pedigrees. We did not seek to create a chronological sequence of objects acquired from 1770s through the 1990s. We did not choose objects to illustrate particular themes such as optics, French makers, or instruments made of wood. Instead, we photographed objects from each disciplinary area of the collection and arranged the pictures into general functional clusters. The sequence of entries also occasionally clusters instruments that had been obtained for different purposes but were used together, illustrating interdependencies among objects that emerged in the astronomical observatory or the lecture demonstration classroom. We did not sort objects into more conventional categories.

For each featured instrument, this book offers a brief photo essay, sometimes including reproductions of related documents. We also provide a brief biography for each instrument, telling stories of instrument-making and design, operation of the device, pedagogy and research, history of science and technology, and local Dartmouth history. We hope that both the sequence and content of our entries will provoke readers to think in new ways about historic scientific instruments and the material culture of past science.

Captions identify each instrument's maker and location (if known), its estimated date of origin, dimensions, particularities of provenance and accession number in the King Collection. Makers' names are rendered exactly as they appear on the instrument. Any text directly transcribed from inscriptions on an instrument is placed in *italics*. Cross references link to other entries in the book.

1

Portable Model L calculating machine – "Executive Monroe"

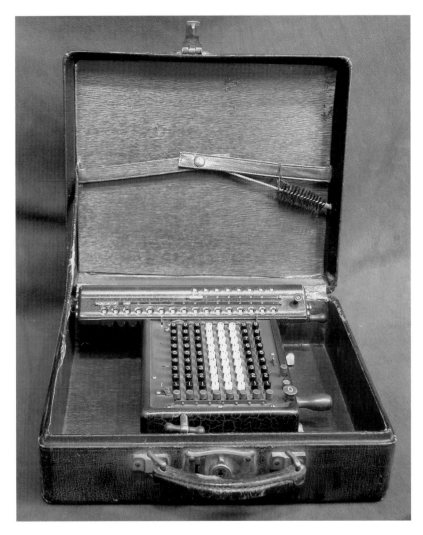

Monroe Calculating Machine Company
New York, New York
About 1933-34
Case: 10 x 27 x 32 cm
Serial number: 163790
Gift of Charles N. Haskins.   2002.1.35224

This manually operated calculator came from the estate of Dartmouth Chandler Professor of Mathematics Charles N. Haskins. During the First World War, Haskins served as "master computer" in the U.S. Army Ordnance Department.

The serial number of this machine dates it to 1933-34.[1] The machine has a baked crackle finish; the case is faux alligator.

Calculating machines such as this one can perform multiplication, division, addition and subtraction.

According to a contemporary review, "The Executive Monroe has been designed particularly for the business man who has personal figure work. Fitting snugly into its own new type, rigid brief case which provides a generous space for papers, it is easily carried so that figure work may be done anywhere—on the train, at home, or out on a field job. It also fills the need of the contractor, engineer, architect, and travelling auditor."[2] Electric calculating machines, emerging at this time, also were marketed for the "busy executive."

Millionaire calculator

Nameplate missing (Hans W. Egli)
Zurich, Switzerland
About 1910
16 x 28 x 76 cm
Serial number: 2688
Gift of George R. Stibitz, Professor of Physiology Emeritus, Dartmouth Medical School.   2002.1.35252

Patented in 1895 by Otto Steiger of Munich and produced and marketed by Hans W. Egli of Zurich, the Millionaire was the first commercially available calculator that could perform direct multiplication with one turn of the crank. Egli sold 4655 Millionaires through the 1930s.

While working as a research mathematician at Bell Telephone Laboratories from 1930 to 1941, George Stibitz developed designs for an electrical "automated calculator." At the Dartmouth Medical School since 1964, Stibitz pursued the application of computer technology to biomedical research.

This Millionaire has been modified by removing the case and replacing part of the top with plexiglass, presumably to make the machine transparent for demonstrating its mechanism. Although convenient for multiplication, the sheer weight of this machine (30 kg) and the necessity of applying numbers from an auxiliary table in order to divide restricted its utility.

THACHER CYLINDRICAL SLIDE RULE

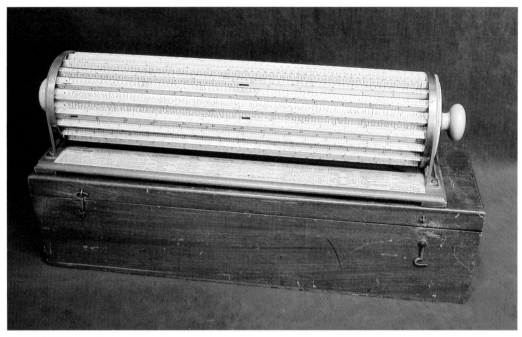

Keuffel & Esser Co
New York, New York
About 1914
14 x 54 x 14 cm
Serial number: 3348

2002.1.35259

William Oughtred, in the early seventeenth century, created a sliding scale with logarithmic graduations. This permitted division and multiplication by the relative ease of sliding the scales to either add or subtract lengths proportional to their logarithms. The number of places to which a linear slide rule can be read depends upon the accuracy of the engraving and the length of the rule. There is a practical limit to the size of a simple linear rule, so many alternative ways of representing the scales were devised. Whereas Oughtred had designed a circular rule, and more exotic types such as helical, where the scales were wound around a cylinder as a helix, giving them greater length, were devised over the ensuing centuries, the linear slide rule (slip-stick) remained dominant to the end of the side rule age.

Charles A. Holden (Dartmouth 1895, Thayer 1900), who taught at Dartmouth's Thayer School of Engineering from 1901 to 1937, strongly promoted the use of calculating machines in American science education. He was particularly enthusiastic about the Mannheim slide rule that would later become the standard calculating device for students and researchers. In 1901 he conducted an extensive survey on the "uses of calculating machines by

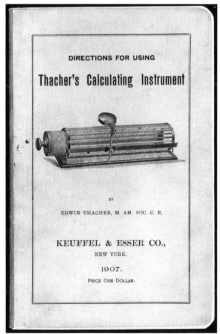

Thacher's instruction manual

5

educators and engineers." One professor wrote to him, "In my judgement, a technical school has not done its full duty by its students unless they have been so drilled in the use of the slide-rule that they turn to it on all occasions." Business firms, he argued, were already aware of the benefit of these machines. "In one case computations on accounts were done with a Thacher machine, doing in three or four days what had usually taken three weeks."[3]

The Thacher cylindrical slide rule was an American contribution to calculating, invented in 1881 by Edwin Thacher, a civil engineer in Pittsburgh, Pennsylvania. Thacher originally arranged for his rule to be produced by the W.F. Stanley Company of London. Around 1887 he transferred commercial development to Keuffel & Esser. Rules made through about 1910 continue to bear the inscription "Divided by W.F. Stanley, London, 1882." The 1907 manual, repeating verbatim Thacher's 1884 edition, indicates that Stanley had developed a special dividing engine to make the scales. "The diamond point used for cutting the divisions [was] moved by a screw of 50 threads/inch, reading by micrometer to 0.000001 inch." The manual further notes that "no time, pains, or expense having been spared to make the instrument in every way reliable; the scales...will give results correctly to four, and usually to five, places of figures, sufficient to satisfy nearly every requirement of the professional and business man." Thacher clearly wished to intimate extreme precision.[4]

Stanley also inscribed the cylinder "Patented by Edwin Thatcher, C.E. Nov. 1st 1881," but spelled the name correctly on the base. Both versions of the inventor's name appear on all extant rules of this design. The machine remained in production between 1882 and the late 1940s.

With twenty triangular segments containing a total of two complete logarithmic scales the effective length is 18 meters. In his manual, Thacher emphasized the practical aspects of the machine, and tried to make the instrument attractive to many trades including accounting, an area where the hand-held Mannheim-type rule was not useful. The Thacher manual contains directions for use in timber measurement, but unlike earlier methods of timber measure (for example, Humphrey Cole's use of Digge's scales), there are no scales specific for this purpose. Cask gauging is also one of the described functions, but gallonage is calculated rather than read off a dedicated scale. In 1907, this Thacher and its mahogany case cost $35. An optional magnifying glass system, illustrated on the cover of the manual, could be obtained for an additional $10. This exemplar was used originally in Dartmouth's Thayer School of Engineering.

At the back of the 1907 instruction manual, K&E advertised their mechanical calculator for a price of nearly ten times that of the Thacher. They were clearly looking toward the future.

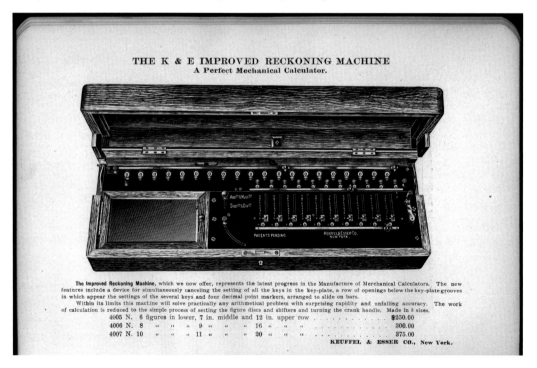

Glassblower's torch; Bunsen burners

Glassblower's torch (left)
Unsigned (Germany)
1880-1920
Height: 20 cm   2002.1.34181

Bunsen burners (left to right)
CENCO
Chicago, Illinois
1900-1940
Height: 16 cm   2002.1.34189

Unsigned
1880-1940
Height: 16.5 cm   2002.1.34602

[L.] E KNOTT APPARATUS CO
Boston, Massachusetts
1880-1940
Height: 14.3 cm   2002.1.34191

Unsigned
1880-1940
Height 16.5 cm   2002.1.34190

   The ubiquitous and familiar laboratory gas burners are used to supply heat for many purposes, from warming solutions to softening glass to make it workable. Most fall under the general category of "Bunsen burners," named after the German chemist, Robert Wilhelm Bunsen, who developed and popularized them in the 1850s. The design was based on previous work done by Michael Faraday. They have a characteristic inlet orifice for the gas and lower openings to draw in air for combustion. Bunsen's key advance was the controlled mixing of air and gas. The air inlets are adjustable, often by simply screwing the tube up or down to expose more or less of the air ports.

   Despite their simple appearance, gas burners have to be carefully designed, with the orifices and air flows matched to the type of gas being burned. In use, an amount of air insufficient for complete combustion is allowed to enter at the bottom. If complete combustion occurred here, the flame would "backfire" and go out. Final combustion takes place at the top. The temperature in the flame from a typical Bunsen burner is not uniform, ranging from about 300^0 C to 1500^0 C.

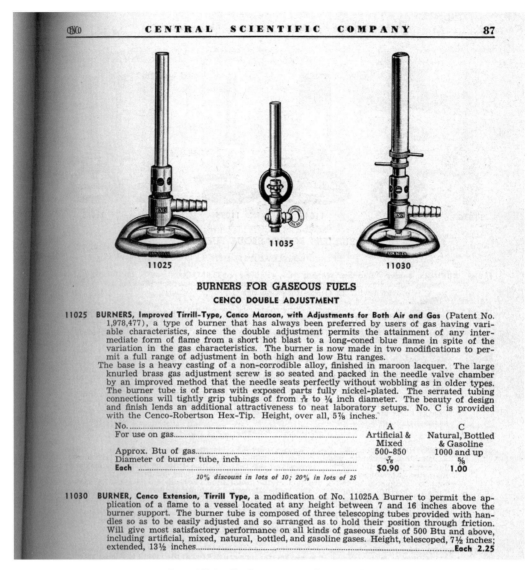

11035

11025   11030

BURNERS FOR GASEOUS FUELS
CENCO DOUBLE ADJUSTMENT

11025   **BURNERS, Improved Tirrill-Type, Cenco Maroon, with Adjustments for Both Air and Gas** (Patent No. 1,978,477), a type of burner that has always been preferred by users of gas having variable characteristics, since the double adjustment permits the attainment of any intermediate form of flame from a short hot blast to a long-coned blue flame in spite of the variation in the gas characteristics. The burner is now made in two modifications to permit a full range of adjustment in both high and low Btu ranges.

The base is a heavy casting of a non-corrodible alloy, finished in maroon lacquer. The large knurled brass gas adjustment screw is so seated and packed in the needle valve chamber by an improved method that the needle seats perfectly without wobbling as in older types. The burner tube is of brass with exposed parts fully nickel-plated. The serrated tubing connections will tightly grip tubings of from $\frac{3}{16}$ to $\frac{1}{4}$ inch diameter. The beauty of design and finish lends an additional attractiveness to neat laboratory setups. No. C is provided with the Cenco-Robertson Hex-Tip. Height, over all, 5⅞ inches.

No.	A	C
For use on gas	Artificial & Mixed	Natural, Bottled & Gasoline
Approx. Btu of gas	500-850	1000 and up
Diameter of burner tube, inch	$\frac{7}{16}$	$\frac{5}{8}$
Each	**$0.90**	**1.00**

10% discount in lots of 10; 20% in lots of 25

11030   **BURNER, Cenco Extension, Tirrill Type,** a modification of No. 11025A Burner to permit the application of a flame to a vessel located at any height between 7 and 16 inches above the burner support. The burner tube is composed of three telescoping tubes provided with handles so as to be easily adjusted and so arranged as to hold their position through friction. Will give most satisfactory performance on all kinds of gaseous fuels of 500 Btu and above, including artificial, mixed, natural, bottled, and gasoline gases. Height, telescoped, 7½ inches; extended, 13½ inches..**Each 2.25**

Central Scientific Company catalogue, 1936

The burner illustrated in the far left (page 7) is designed to be a blast burner; gas and air are forced in at high velocity and create a higher heat than possible with the conventional Bunsen arrangement. Such a blast burner is used to work higher-melting-point glasses, such as Pyrex, or to melt small quantities of metal. This burner is of German origin, as evidenced by the inlets marked *Luft* and *Gas*. Its two ball joints give it directional flexibility. The burner to its right, with the cup at the top, has the top of the tube wrapped with an asbestos wick. This was evidently used to produce a flame with spectral characteristics of the solution placed in the cup. Its base is baked maroon lacquer.

The hose connections on these burners show a variety of serration patterns. The ones on the blast burner are quite robust, presumably to hold the tubing against pressure, while the other burners have more modest grooves.

The illustration from the 1936 Central Scientific catalogue shows only a few of their burners. Note the large-quantity discount offered, suggesting that these burners were supplied to schools and industry in great numbers. In addition, Central Scientific designed their burners so "that graceful contour and beauty of finish be such as to give attractiveness to laboratory setups employing the burners."[5]

8

PRECISION THERMOMETERS

C. Gerhardt
Bonn, Germany
About 1885
18 x 9 cm

2002.1.34227

   This set of five precision glass thermometers, in a fitted velvet-lined case, has a range of -10° to +260° C, marked in quarter degrees.

   Founded in 1846 as Marquarts Lager chemischer Utensilien, the company was purchased in 1872 by Carl Gerhardt and renamed C. Gerhardt Fabrik & Lager chemischer Apparate, the name maintained to the present. Gerhardt is thus one of the oldest instrument companies still in existence.

CUCURBIT AND HEAD

Unsigned
Late eighteenth century
Height: 38 cm; Diameter: 25 cm

Exhibited at the Joseph Priestley House, Northumberland, Pennsylvania, 1976-1985.   2002.1.35312

Similar to a retort, this hand-wrought copper vessel was used for distillation. The apparatus consists of three parts–the snout, the head and the main body (cucurbit). They are held together with lute, a clay-like refractory cement used for this purpose. The seams are riveted.

Instruction in chemistry began early at Dartmouth. Lyman Spalding, who had studied medicine at Harvard and Dartmouth, lectured on chemistry at the Dartmouth Medical School from 1798 to 1800. Spalding was the second American author to write a separate work on the new chemical nomenclature of Lavoisier, publishing in 1799 a reworked table addressed "to the students of chemistry at Dartmouth College." A reviewer in the *Medical Repository* lauded Spalding's text and "the taste for chemical inquiry [that] has manifested itself at Dartmouth College." In 1805, the Harvard professor of chemistry enviously noted that Dartmouth had appropriated $600 for chemical apparatus. From 1813, Dartmouth's prescribed curriculum for the junior year included chemistry, natural philosophy and astronomy, and William Paley's *Natural Theology*.[6]

Despite this early activity, the King Collection contains very little apparatus from chemistry. The earliest inventory of chemical apparatus, prepared in 1815, lists about thirty items, none of which are extant except for the "cucurbit with head for Subliming."[7]

Mann's short-interval timer

MANN
Hanover, New Hampshire
1924
6 x 11.5 x 11.5 cm

2002.1.34752

This timer is essentially a precision stopwatch for laboratory use, measuring intervals of time up to two minutes, in units of 0.1 seconds. David W. Mann designed and built this mechanical timer while at Dartmouth, and secured its patent in 1924. Soon thereafter, Mann moved to Cambridge, Massachusetts and established the Mann Instrument Company. Leaving that firm in 1936, he subsequently founded the David W. Mann Company of Lincoln, Massachusetts, which produced precision motion devices used in ruling engines, astronomical comparators, and the early manufacture, in the late 1950s and early 1960s, of integrated circuits.[8]

Spring-driven, bench-top interval timers became a standard item in laboratories. Later devices often include an audible alarm to signal the end of a selected interval of time.

King's thermoelastometer

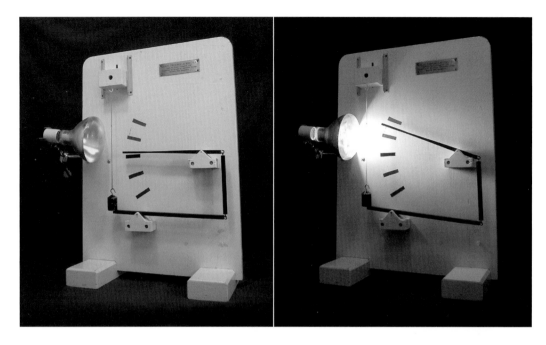

Unsigned (Allen King and Ernest Fitzgerald)
Hanover, New Hampshire
1948
63 x 46 cm   2002.1.35418

 This thermoelastometer, used for teaching purposes, was designed by Professor Allen King at Dartmouth in 1948, and was built by Ernest Fitzgerald, a technician from 1937 to 1959 in the Physics Department. It illustrates how "homemade" instruments often are constructed to satisfy a need that commercial makers do not fill.

 King designed the device to demonstrate the unusual thermal properties of rubber-like substances, called Gough-Joule effects. A thin rubber band supports a weight that rests on the short end of a compound lever system. When the heat lamp is turned on, the rubber, a polymeric substance, contracts, and the compound lever amplifies the motion so that it is clearly demonstrated on the arbitrary scale. At this time King was studying the elastic properties of human tissue.[9] Many constituents of connective tissue are polymers.

Dip circle

Phelps & Gurley
Troy, New York
About 1848
Height: 38 cm; Diameter: 25 cm   2002.1.35225

The dip circle (sometimes called an inclinometer or dip needle) is essentially a vertical compass needle, used to determine the vertical component of the earth's magnetic field. In 1576, the English hydrographer and instrument maker, Robert Norman, invented the device and reported the first systematic measurements of dip. William Gilbert in 1600 proposed that the earth was, in effect, a magnet, and illustrated how the dip needle would change at different latitudes. Others suggested that this phenomenon could be used to determine latitude. However, despite many attempts, particularly those of Edmund Halley, geomagnetism was not successfully used to determine either latitude or longitude until the development of the earth inductor compass early in the twentieth century.

William Gurley and Jonas Phelps had a short-lived partnership in Troy, New York, producing surveying instruments. Lewis Gurley also worked for the firm. In 1852, the Gurley brothers assumed ownership, and began trading as W. & L.E. Gurley. This firm became one of the largest manufacturers of engineering and surveying instruments in the United States. On a purchasing trip to Europe in 1875, Professor John Foster of Union College noted with pride that "there is not in England or on the continent a workshop equal to that of Gurley's at Troy."[10]

The horizontal arm holding the needle bearings is signed *Phelps & Gurley*.

The bearings supporting the magnetized needle are simple and effective. The magnetized steel needle has a small axle that simply rolls on flat agate slabs. When not in use, the needle is lifted off the agate by means of an arm actuated by an external knob. The act of lifting the needle restores the axle to the center of the agate, negating any distance it might have rolled.

The rotation of the horizontal circle can be controlled by a knob attached to a vernier, reading against the outer ring to tenths of an arc degree, giving the azimuth.

VARIFLUX MAGNET

LABORATORY FOR SCIENCE
Oakland, California
About 1959
28.5 x 35 cm

2002.1.35165

The Variflux is a permanent magnet designed for demonstration purposes. The magnetic flux can be varied over a wide range; there are well-designed adjustments and the settings are repeatable. The "blue-gray hammertone" finish was frequently seen on laboratory instruments in the 1950s and 1960s.

Variflux was a wholly-owned trademark registered to Laboratory for Science from 1959 to 2001.

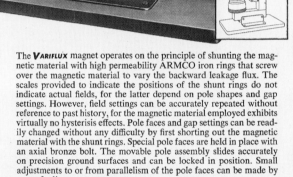
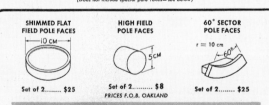
Variflux literature

16

EIGHTEENTH-CENTURY MECHANICAL POWERS

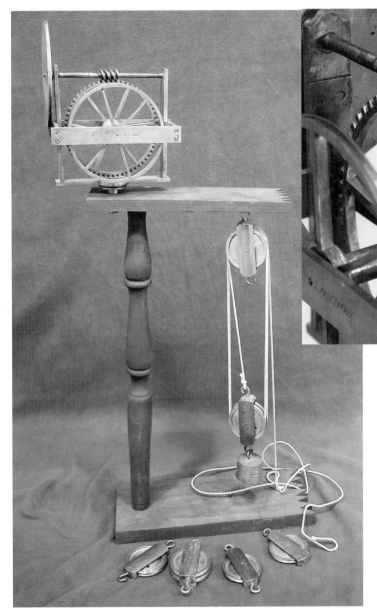

GILBERT
London, England
1785
Height: 59 cm

2002.1.35200

In 1787 Dartmouth College had two full sets of mechanical powers, one of them "elegant."[11] They were used to demonstrate to students how wheels, pulleys and gears transfer power. This is all that remains of one set. The turned wooden stand is later. There is physical evidence of repair, and possible replacement parts.

The apparatus is made of brass, but the 3-lead worm screw is iron. It meshes with a brass gear that has teeth hobbed at an angle, but flat on top, not concave. The worm gear has a sliding clutch (seen near the hub on the horizontal spoke in the foreground in the detail above) that disengages the worm gear assembly from the pulley, which then free-wheels.

One of London's leading instrument makers during the second half of the eighteenth century, John Gilbert, Jr. is known to have sold a variety of mathematical and optical apparatus.

A similar mechanism, seen below, is illustrated in George Adams' *Lectures on Natural and Experimental Philosophy* (1807). Adams' mechanism is shown with a single-lead worm running against a 48-tooth wheel. Dartmouth's Gilbert mechanism, with a 3-lead worm running against a 60-tooth wheel, requires 20 turns of the flywheel (connected to the worm) to rotate the wheel once, giving a 1:20 mechanical advantage. The advantage is 1:48 in the machine illustrated by Adams. There is no evidence that the machine in the engraving has a clutch to disengage the worm from the rest of the apparatus. From 1806, Dartmouth's curriculum for the junior class mandated the study of Adam's *Lectures*.

There is a mechanical powers in the British Museum, signed by Adams, that has a design identical to ours, including a 60-tooth wheel and sliding clutch. This suggests that London makers reproduced each others' work.[12]

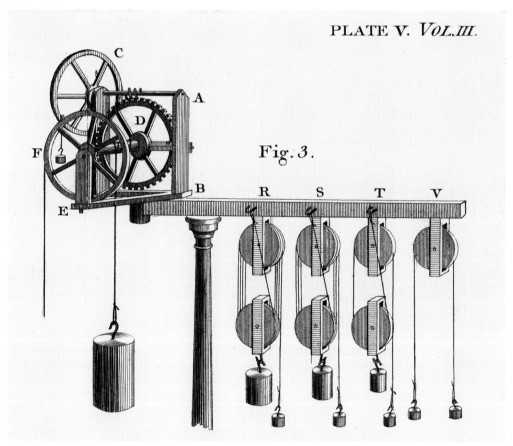

Copperplate engraving, Adams' *Lectures,* 1807

SEISMOGRAPH

Unsigned (designed by Louis Bell)
Boston, Massachusetts
About 1904
Height: 27 cm

2002.1.35226

Louis Bell (Dartmouth College, 1884) received his PhD in physics and electrical engineering in 1888 at Henry Rowland's laboratory, Johns Hopkins University. He was one of Rowland's first doctoral students and did pioneering work with Rowland's new diffraction gratings (see page 95).

Bell taught briefly at Purdue University and moved to Chicago, where he was a consulting engineer with Bliss & Bell for about a year. He then went to New York to become Editor of *Electrical World*. A few years later he joined General Electric Company as their chief engineer of the Power Transmission Department. In 1893, Bell designed and installed the first 3-phase electrical transmission plant. The next year he left GE to open an office in Boston as a consulting engineer. He received over forty patents and wrote what became classic treatises on electric power transmission and illumination. He had a lifelong interest in astronomy, writing in 1922 *The Telescope*, a widely popular book.

Bell became interested in the vibration of the earth when, at the time of the Spanish-American War, he was the technical officer in charge of mining Boston Harbor.

Shortly after the war he built several seismographs, one to register horizontal movement and one to register vertical movement. He used these to determine the effects of outside shocks on buildings, some of which resulted from street cars, trucks, and trains, both elevated and running in tunnels. He also measured the effects of blasting.[13]

19

Bell's seismographs operate on well-known physical principles, using a pendular mass that tends to remain stationary when the rest of the instrument is displaced by seismic movement. The deflection is recorded on a smoked paper disc.

The instrument has a fitted wooden box that contains a small wooden carrier that holds up to ten pre-smoked discs. Clearly, the instrument was meant to be portable. There is also a square brass plate with a central hole, beveled on one side, that forms a template for cutting the discs. Judging from the presence of random bits of text on the underside of the paper, Bell recycled used paper for his smoked discs.

The instrument, a custom-made piece, must have been made by a master machinist. The surfaces of the heavy brass upper and lower rings are decorated with a curious swirling pattern that appears to have been made by a scraper. Since there is no obvious reason, related to building this device or its function, for these brass surfaces to have been scraped, a possible explanation is that the scraping patterns create a kind of

"signature." Many machinists developed a characteristic motion when they scraped, and the final pattern, or frosting, can be unique. The small inset to the right illustrates frosting on a steel plate scraped by a master scraper in the early part of the twenty-first century.

Machine-tool rebuilder Edward Connelly notes: "No one else will be able to imitate his mark exactly and it will be as personal as handwriting .... Machinery bearing this mark would be identifiable throughout the world."[14] For another example of such scraping, see the Zentmayer microscope entry (page 104). The King Collection also has a Crouch microscope (2002.1.34296) with similar decorative frosting on the stage.

20

GRAPHOMETER AND GUNTER HALF CHAIN

Graphometer
Unsigned (William Guyse Hagger?)
About 1770
21 x 11.5 cm   2002.1.35211

Gunter half chain
Unsigned
About 1770
Length: 11 yards   2002.1.35227

By the eighteenth century, surveying a plot of land required measuring angles and distances in the field, plotting these data on paper, and calculating the area. The chain is used to measure linear distance. In 1620, Edmund Gunter, professor of astronomy at Gresham College in London, devised an iron-link chain for this purpose. A full length Gunter's chain is four rods (22 yards) in 100 links; four rods squared equals one-tenth acre. The chain shown here is half this length; half chains were needed for areas of dense undergrowth, which prevented full length chains from being stretched taught. Handles on the end of the chain make it easier for two individuals to pull it straight.

The graphometer, developed in the early eighteenth century, is a surveying instrument with a magnetic compass, an alidade (rotating sight) and a divided half-circle. It is a simple and relatively cheap instrument used to measure angles. By laying out a triangle with one known side, an area can be calculated. Such trigonometric surveying practice replaced earlier methods wherein areas were laid out in measured rectangles, and the area of odd-shaped remnants estimated.

The entire device is rough, perhaps revealing rustic origins. Silvio Bedini suggested that this graphometer may have been made by William Guyse Hagger (1748-1832) of Newport, Rhode Island. Bedini also illus-

trates a wooden graphometer signed by Hagger's son, Benjamin King Hagger, that is very similar in design to Dartmouth's.[15] American wooden surveying instruments are relatively scarce and seem to have been made largely in New England during the eighteenth and early nineteenth centuries.

The brass sheet, nailed to a wooden slab, is engraved with a half-circle of degrees and is no longer securely affixed to the wood. The alidade sights were not designed to fold; one shows solder repair. The compass (now inoperative) is free to move a few degrees to either side of north. A spirit level is set into a roughly gouged-out cavity on one side of the wood, held in place by a thin brass plate. The graphometer is designed to measure both azimuth and altitude. In the center of the bottom is a brass fixture to hold the

graphometer to the tripod or pole. Part of the fixture seems to be a new replacement. Its position was laid out by scribed marks from each corner. Still visible, they suggest the wood was not finished before assembly, further contributing to the rustic appearance of this piece.

The wood appears to be recycled and shows many marks and cuts that are not readily understandable in terms of this device. There is, for example, next to the spirit level, a small indented circle with points. A rectangle has been cut out of the spirit level's brass retaining plate and the wood underneath gouged out. There is a similar hole in the wood, visible from the underside, that may once have had a brass plate affixed over it. The two brass tabs on the underside appear to have no function.

This chain and graphometer are reputed to have been used to survey the original College lands by the founder, Eleazar Wheelock, and his assistant, Bezaleel Woodward, the first professor of mathematics and natural philosophy. Woodward's classes emphasized practical problems, especially those related to surveying. From at least 1796 (the earliest formal description of the curriculum by the trustees) through 1879, all Dartmouth undergraduates were required to take a course on surveying. In the 1870s, the surveying students were offered "the use of the instruments and practice in the field, drawing of plans and maps, triangulation."[16]

Surveyor's quadrant

John Kennard
Newmarket, New Hampshire
About 1820
Radius: 19.5 cm
Gift of Frank C. and Clara G. Churchill of Lebanon, New Hampshire, 1946.
Exhibited in "Illuminating Instruments," Hood Museum of Art, Dartmouth College, 2004.   2002.1.35217

   This mahogany and brass instrument, in the original fitted pine box, is a product of three little-known New Hampshire craftsmen. The brass sides are inscribed *Invented by P. Merrill, Esq.* and *Made by John Kennard Newmarket*. The compass card, printed on thick paper in red and black ink, is signed (twice) *T S BOWLES PORTSMOUTH NH*. Three other nearly identical quadrants, similarly inscribed with these three names, are known. Two are in private hands and one is in the Currier Gallery of Art, Manchester, New Hampshire.[17]

The three men whose names appear on the quadrant lived in communities near the mouth of the Piscataqua River, in 1800 still a major New England seaport and center of trade and commerce. A member of Portsmouth's seafaring Salter family, Thomas Salter Bowles by 1806 was advertising the availability of "azimuth and brass compasses, wood and hanging compasses" at his shop in that town. The earliest directory for Portsmouth (1821) lists Bowles as a "mathematical instrument maker." He moved to Portland, Maine, in 1825. In both states, Bowles held high office in Masonic lodges. The Masonic emblems of the builder's square and compass appear prominently below the fleur-de-lis on Bowles' compass cards. He placed the American eagle on the east compass point. The compass card is divided into 32 points and 360 degrees. The colorful card, rippled glass (cracked) and putty all seem original.

Phinehas Merrill (1767-1815) of Stratham, New Hampshire, was widely known as a surveyor, school-teacher, and member of the state legislature. For several years, he surveyed towns for the first map of New Hampshire based on measured surveys. Exactly what Merrill "invented" on these quadrants remains unclear. His name is not known to appear on any other extant surveying instruments from the period.

John Kennard (1782-1861) had learned the trade of clockmaking in Portsmouth before moving to nearby Newmarket in 1812. He kept store, served as postmaster, as was especially known for his tallcase and banjo clocks. Apart from these four quadrants, no other surveying instruments made by Kennard are known. The division of labor implied by the inscriptions on this quadrant suggests the existence of subspecialties in the Northern New England instrument trade during the early national period.[18]

The quadrant can be positioned to measure angles of azimuth or altitude. The alidade has folding sights and a vernier. The quadrant is graduated to thirds of a degree; with the vernier angles can be marked to one-tenth of a division or two arc-minutes. A freely swinging brass bob is given to establish verticality, with a benchmark scribed into the wood under the pointer. On the back, a brass collar is screwed onto an inset brass plate. A brass rod, terminating in a ball, fits this collar and is held in place by means of a thumbscrew that fits into a groove on the rod. All of this appears to be original. The ball fits into a two-part brass fixture that appears to be modern. When the lower part of the fixture is screwed tight, it bears on a cork, locking the ball in place by friction. This device works remarkably well, permitting easy movement, and clamping does not change the position of the ball. The entire device would fit the top of a staff or tripod.

A small round brass plate on the back swings open to reveal a locking mechanism for the compass needle.

The wood and brass components appear to have been made from salvaged material. The brass bob may have been fashioned from scrap.[19] The three slots visible on the top surface of the quadrant, along one of its straight edges, were designed to hold a pair of removable brass sights that could be inserted to add a fixed sightline to the quadrant. When not in use, these extra sights (missing in our instrument) would have been stored in a chamber hollowed out of the edge of the mahogany and covered with a brass door (visible at the bottom of the illustration on page 25).

SURVEYOR'S COMPASS (CIRCUMFERENTOR)

Unsigned
About 1800
Length: 37 cm; Compass diameter: 16 cm

2002.1.35210

The surveyor's compass, consisting of a magnetic needle and sights so that bearings with reference to the magnetic meridian can be taken, found its earliest use in colonial settings such as Ireland and America where vast lands awaited measurement. The term "circumferentor" was used primarily in England. By 1750 in North America, the term "surveyor's compass" was preferred. In addition to surveying wilderness, these compasses were also used to lay out mine tunnels.

On North American instruments, such as this one, the compass card typically displays west to the right of north. In this arrangement, when the user sights from the south to the north end of the device, the needle registers the direction (azimuth) he or she is looking. The card moves with the alidade, the needle remains stationary. The theodolite, in contrast, has a fixed divided circle and an alidade acting as the pointer, which rotates independently of the circle.

The compass, with a removable glass cover, is graduated in degrees around its inner rim. It lacks a compass card or, as is more generally the case on brass instruments, compass points engraved directly upon the metal face. It has two spirit levels perpendicular to each other. The sights attach by thumb-screws. The wooden cover fits over the compass and is tied with string. Tool marks on the interior of the cover show that the cavity was partly turned and the walls were finished by hand, using a gouge.

Literally hundreds of eighteenth- and nineteenth-century surveyor's compasses, produced by dozens of American makers, are extant in museums and private collections across North America. In 1820 one Philadelphia maker, Thomas Whitney, boasted that he had made more than five hundred compasses, "the good qualities of which are known to many Surveyors, in at least sixteen of the States and Territories of the Union."[20]

SURVEYOR'S COMPASS (CIRCUMFERENTOR)

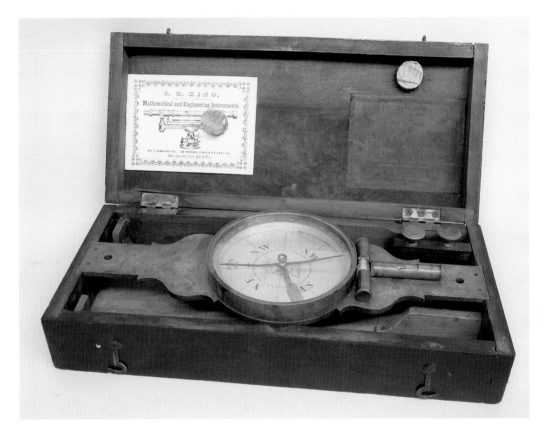

H.M. Pool
Easton, Massachusetts
Date uncertain, 1841-1880
Length: about 35.5 cm; Compass diameter: 12 cm

2002.1.35408

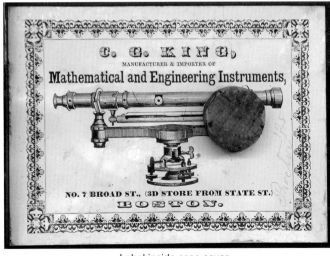

Label inside case cover

Horace Minot Pool started making instruments with his older brother, John, in the 1820s, and went into business for himself in 1841. In 1860, it was reported in the Federal Census of Industry that Pool had made 55 compasses worth a total of $1650, plus transits and theodolites worth another $2850.[21]

The wooden case displays two additional labels. Charles Gedney King, after apprenticing with his father, began in 1838 to manufacture and import mathematical instruments under his own name.

29

The less legible label on the right is for Thaxter & Son, another Boston firm founded in 1822 by Samuel Thaxter and his son, Joseph. Thaxter & Son billed itself as an importer and dealer. Pool's compass fits snugly into the case bearing labels by King and Thaxter & Son. Without further documentation, one can only speculate about the relationships among these three makers that may be implied by this compass in its case.[22]

Pool's instrument has two horizontal spirit levels at right angles to each other. Its sights, held in place by thumbscrews and aligned by pins, are removable and thus allow the instrument to fit within its flat case.

Optical square

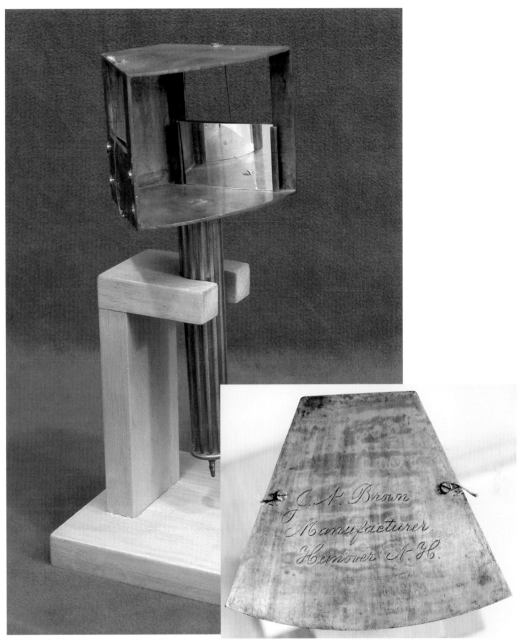

J.N. Brown
Hanover, New Hampshire
About 1885
Height: 22 cm

2002.1.35213

An optical square is used for sighting along two lines at right angles to one another, particularly in surveying. For a discussion of John N. Brown, the local maker, see entry for the tangent galvanometer (page 175). The mirrors have been replaced in this instrument.

31

WYE LEVEL

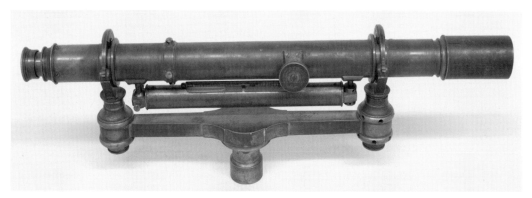

Wm J Young
Philadelphia, Pennsylvania
About 1846-53
Length: about 44 cm

2002.1.35409

   This brass instrument permits the establishment of a level line across distance. It is robust and able to hold adjustments in the field. Levels such as these were used widely in laying out grades for canals and rail beds. Focusing is accomplished by moving the objective lens that is coupled to the focusing knob by a rack and pinion. The instrument is missing its reticle or cross-hair. It gives an erect image. Adjustments are provided to align both the level and the base with the telescope. Holes in the end-ring support and the nearby level support (capstans) serve to accept spanners or rods that can be used to turn them, effecting the height alignments.[23]

   The instrument was also referred to as a Y level. Its design remained in use for a long time, and was manufactured by several makers.

   William Young was the first American maker to own a dividing engine. He had a thriving workshop with as many as ten workers and apprentices, some of whom were highly skilled and demanded high wages. Young's instruments were therefore quite costly.[24] Starting about 1853, Young began to place serial numbers on all his instruments. This one lacks a serial number.

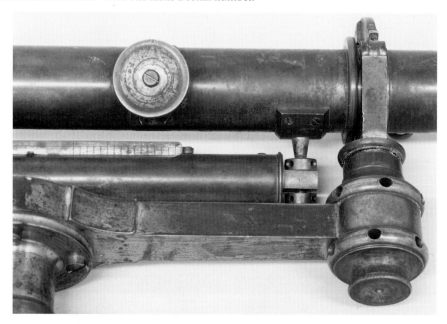

SUNDIAL

Heath & Wing
London, England
1773
Diameter: 52 cm
Gift of Samuel Holland, 1773.
Exhibited in "Illuminating Instruments," Hood Museum of Art, Dartmouth College, 2004.   2002.1.34750

"There is a very fine brass horizontal dial fixed on a post in the President's yard. It was given by Capt. Holland; it cost ten guineas. The latitude of the place: 43° 38' N," reported a visitor to Dartmouth in 1774. The donor, Samuel Holland, was surveyor general of the Northern District of North America during the late colonial period. Between 1764 and 1774 he conducted a survey of the coast from Prince Edward Island to Boston. In the early 1770s his crew was stationed at Portsmouth, New Hampshire. Holland was a good friend of John Wentworth, provincial governor, who took him on a rugged inland journey to attend Dartmouth's second commencement in 1772 (two graduates). In 1777 Dartmouth's first professor, John Smith, was hired at an annual salary of £100. A sundial costing 10 guineas (£10 10s) thus represented a gift that signified confidence in the new institution.[25]

This dial is the earliest signed and dated instrument now in the King Collection. For over half a century, it stood on a post in the president's yard. It then disappeared into the Shattuck Observatory, where it was found in 1929 and removed to Baker Library. In 1967 Allen King "rediscovered" the dial, and in seeking a prototype for its missing gnomon, located two similar Heath & Wing dials then held by a London antiques dealer. The last Heath & Wing catalogue (1771) offers "horizontal sun dials, of various kinds."[26]

The original gnomon, possibly lost early in the nineteenth century, had been replaced by a simple triangular piece of wood found with the dial. The current brass gnomon, made in 1985 by Richard C. Johnson, manager of the Dartmouth College Apparatus Shop, appears unusually thick, but horizontal

dials of the period typically had thick gnomons supporting the shadow-casting style and screwed from underneath to prevent theft or vandalism. The thickness in effect gives the gnomon two styles; the hours on our dial are laid out so that the morning and afternoon shadows are correct.

Long considered one of the central icons of natural philosophy at Dartmouth, the sundial has featured prominently in histories of the College. Yet the chroniclers have not agreed on its composition, some calling it "brass," others "bronze." Eighteenth-century use of the word "brass" does not necessarily equate to present-day use of the term, referring to an alloy of copper and zinc. Bronze today is considered to be an

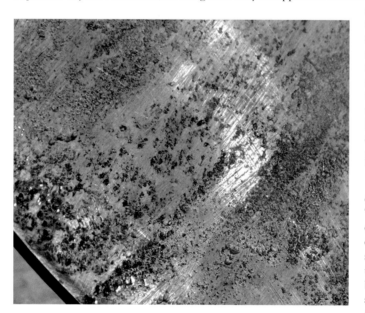

alloy principally of copper and tin, but many eighteenth-century artifacts denoted as bronze contain a significant amount of zinc. Samuel Johnson's dictionary of 1775 defines brass as "a yellow metal, made by mixing copper with lapis calaminaris. It is used, in popular language, for any kind of metal in which copper has a part." Bronze he defines as "relief or statue cast in brass."[27]

This sundial has the characteristics of a cast piece. The back shows pits and fissures characteristic of metal allowed to cool in an open mold. We note also the many grinding marks on the back. Although these could be later, they suggest that the slag and rough surface which formed on the exposed surface was ground off to partially smooth it. Had the disc been cast in a closed mold (such as a cope and drag), such a surface would not be expected. Whereas the top appears to have been patinated, the bottom was not.

34

MARINE CHRONOMETER

Margetts
London, England
About 1795
Diameter: 14 cm
Serial number: 83

2002.1.34751

George Margetts, one of the famous second generation of pioneer chronometer makers, was a skilled mathematician who published mathematical tables for the lunar distance method of finding longitude. He also invented a number of peculiar improvements for chronometers, including the side-winding mechanism found on this instrument.

As attested by notations on its inner surfaces, this typical 8-day Margetts chronometer was used on board a ship through the mid-1880s. On the inside of the case, we see *7/7/63* and *XII.* On the back of the brass plate supporting the enamel dial we find *Jesse Smith, Jr./Salem Oct 21, 1825* and *EB/Marseille 1883.* Jesse Smith, Jr. was a well-known clockmaker in Salem, Massachusetts. Judging from the calligraphy we

presume that the 7/7/63 refers to 1863. These notations may record four separate repairs. It has long been an accepted practice for clockmakers to leave their marks after opening and repairing an instrument.

The instrument shows physical evidence of having been opened many times, and possibly of having had the case modified or repaired, even recently. There are stainless steel fillister-head screws attaching the bezel. William Bond & Sons, a leading Boston clockmaker who in the 1830s was repairing and delivering chronometers for U.S. Navy ships in Boston and Portsmouth, entered a record into their day-book on 10 October 1834, indicating work done on this chronometer, including "New ball[ance] spring & repairs.[28]

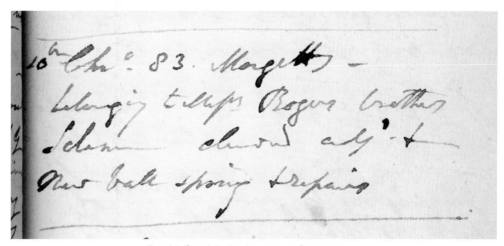

Bond & Sons' daybook entry, 10 October 1834

The physical evidence in conjunction with the entry in the Bond & Sons daybook collectively suggest that Dartmouth acquired this chronometer no earlier than late in the nineteenth century. Presumably the chronometer had spent almost a century aboard ships. Shipboard usage also explains the double gimbal mounting. Margetts generally did not provide gimbals for his instruments; they were meant to rest flat on a table. Hence this gimbal probably is later, perhaps 1800-1810.

MARINER'S OCTANT

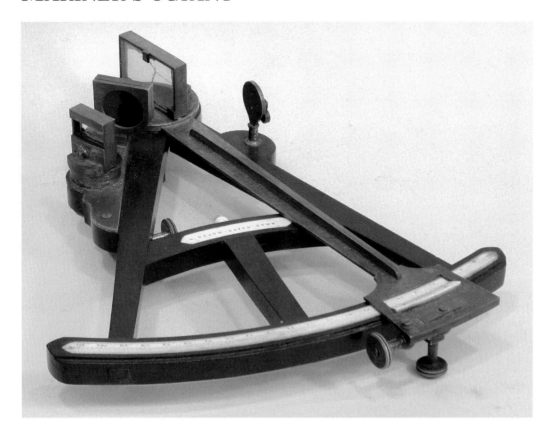

G. Heath
Erith, Kent, England
About 1850
Radius: 24 cm   2002.1.35228

Imported by *Thaxter & Son, importers and Dealers in Nautical and Mathematical Instruments No.125 State St Boston, Sextants, Quadrants, Compasses, Spy glasses, Barometers, thermometers, Surveying and Gauging Instruments, Charts and nautical Books,* according to the printed label in the wooden case (see also Pool's surveyor's compass, page 29).

The scale on an octant is 1/8 of a circle, or 45 degrees. Mirrors, used in the octant to make a star or the sun coincident with the horizon, double the angle, so that altitudes of up to ninety degrees can be measured. A sextant, in contrast, has the scale divided into one-sixth of a circle, or 60 degrees. As in the octant, the mirror doubles the angle, hence the sextant can measure from 0 to 120 degrees. Both instruments were designed early in the eighteenth century for navigation at sea. The greater angular range of the sextant made it ideal for "taking lunars," i.e., determining the angle between the moon and sun or stars for obtaining longitude. Octants were in use until the twentieth century.

This octant is made of ebony with a brass index arm. It has a pinhole sight, solar filters and vernier, and is complete with fitted wooden case. The scale and maker's name are on inlaid ivory.

George Heath was an obscure maker, known primarily for exhibiting sextants at the 1851 Great Exhibition in London.

U.S. Army Air Force bubble sextant

ANSCO A DIVISION OF GENERAL ANALINE & FILM CORP
Binghamton, New York
About 1940
10 x 15 x 15 cm
Exhibited in "Illuminating Instruments," Hood Museum of Art, Dartmouth College, 2004.   2002.1.35375

Aerial celestial navigation presents unique problems. Often a horizon is not visible, or if one is, the difference between the observed horizon and the horizontal (called the dip) becomes very large and, because of altimeter uncertainty, often has an unacceptable error.[29] A leveling device was needed for aerial use of a sextant and most often this was a bubble whose image was coincident with that of the sighting image. This is a Type A-10 sextant, one of several designed specifically for aerial navigation by United States military forces at the time of World War II. The sextants were manufactured by a number of different makers. The navigator took his sight through a plexiglass dome in the plane's fuselage. The bubble could be adjusted in size and illuminated, using a battery-powered lamp, for night use. A recording disc, upon which observations could be marked at one-second intervals, was an integral part of some sextants. An average of the sightings (usually about 60-90) could then be obtained.

FAIRCHILD TYPE A-10

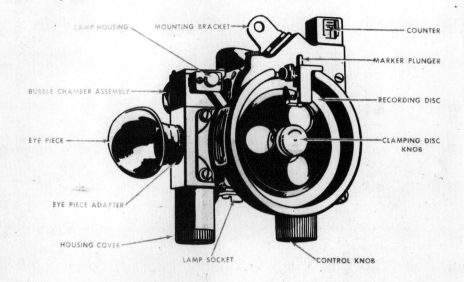

Classified instruction manual

Care and Handling

Keep your fingers off the prisms to avoid getting the glass dirty. Be particularly careful when removing or replacing the instrument in the carrying case. If the prisms do become dirty, clean them with a well-washed, soft handkerchief, or lens paper.

The prism control knob has a turning arc from 0° to 90°; in addition it has approximately 1½° of extra movement above and below these limits. If you attempt to turn it past its limits you are likely to damage the instrument.

When attaching the lamp plug to the instrument, push it straight onto the connection and remove it by pulling it straight off. Don't bend or twist the plug. Bending or twisting will loosen the plug or break the connection inside the housing. In either case the lamp assembly will not work.

Remove the batteries from the battery case when you are not using the sextant. It is a poor policy to carry batteries in the battery chamber merely to have more spares. Even the non-leaking variety of

batteries will cause some corrosion.

In relatively smooth air you can turn the wax disk by loosening the knob holding it, so that you can take several shots using the same disk. If the marks extend over a considerable portion of the disk you will have to replace it before each observation.

Turn the bubble to maximum size when you put the sextant back into the case. To do this, turn the bubble control knob clockwise until you reach the limit of turning. This pulls the diaphragm out, and increases the space in the bubble chamber. By increasing the space (since you have a constant amount of fluid) you get a larger bubble. The reason for putting your sextant away with a large bubble is to allow for temperature changes. If expansion occurs it will be absorbed by the bubble and will not break the diaphragm.

Wax disks melt if subjected to heat, so keep your sextant case away from the heating tube of an airplane. If the case becomes hot, the wax on the disks will melt and the whole supply will stick together and they will all be useless.

During WWII Dartmouth hosted a United States naval training school (V-12 program). Richard H. Goddard, then director of the Shattuck Observatory, taught many of the courses in this program, including those on navigation. After the war, Goddard obtained ten of these A-10 bubble sextants from war surplus, along with their "restricted" instruction manual. Goddard annually offered a course on navigation until he retired in 1963.[30]

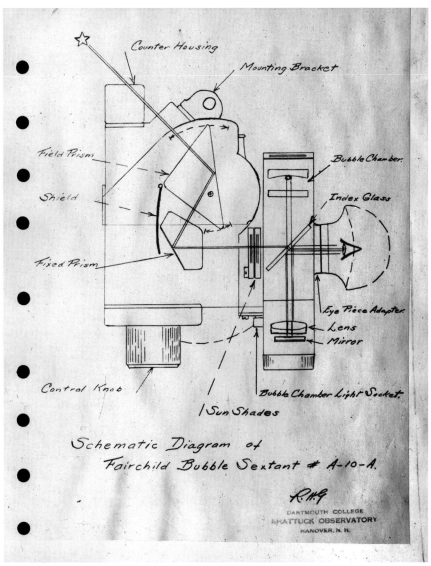

This diagram, bearing the initials of Robert H. Goddard, shows nicely the compact nature of the optical system of the bubble sextant. The center of the bubble indicates true vertical, its margins create a "horizon." As can be seen in the diagram, the optics superimpose the bubble and the star.

Bubble sextants were not limited to military use. They aided navigators on long-distance flights in Zeppelins, and were used in commercial civilian aviation until relatively recent times. Coincident with the development of airplanes suitable for passenger transportation, especially over long distances, several systems of radio and inertial navigation were developed. Part of the pressure to do this came from airlines who wanted to eliminate the highly-paid navigator from the cockpit. This became a reality when Doppler systems were developed; when GPS was installed, the need to know celestial navigation was eliminated completely. Retired TWA Captain Bob Buck summed it up: "Now it's all gone, future fliers will never experience the romance and skill of being guided by the stars. It might seem my learning effort was wasted, but it wasn't; it was necessary for its time, and gave me treasure, something I realize whenever I stand outside on a crystal-clear, cold winter night and look up to my old friends, Sirius, Vega, and Polaris."[31]

41

Japanese sextant

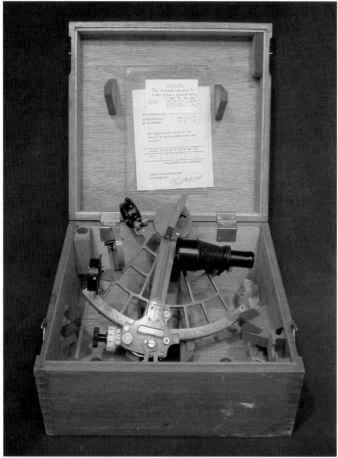

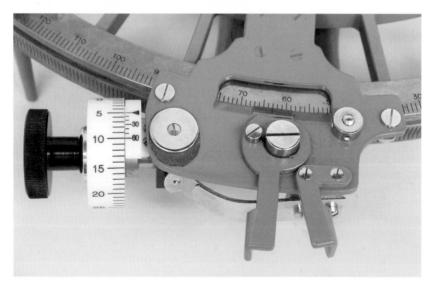

KAGIYA YOKO
Osaka, Japan
About 1966
Radius: 16.5 cm
Gift of Fred Springer-Miller (Dartmouth, 1949), 1994.

2002.1.35395

Label on top of lid

This Japanese bronze sextant was the type used by both Japanese merchant and naval vessels in the first half of the twentieth century. It was presented by the donor because "Dr. Goddard taught me navigation in 1945."[32]

According to the donor, this design has zero index error. Index error results from misalignment of the scale. In actual use, the horizon is sighted and the difference between 0⁰00'00" and the reading obtained is the index error. An instrument with zero index error obviously facilitates observation reduction.

The brass, blue-painted frame of this sextant has been designed to minimize weight, provide rigidity and minimize wind resistance. This well-made instrument is from the twilight of the era of celestial navigation.

ORRERY

Unsigned (*BMC*?)
London, England
About 1780, purchased 1785
Diameter: 32 cm; Height to plate: 14 cm

2002.1.35201

Dartmouth's first order, in 1773, for instruments from London was initially delayed, then cancelled because of the American Revolution. After the war, in 1785 the brothers of John Wheelock (son and successor of Eleazar Wheelock) brought this orrery back from England. From 1778 through the 1840s, astronomy was a required subject for each junior class at Dartmouth. In 1834, Ira Young, newly appointed professor of mathematics and natural philosophy, spent $0.75 on "repairing orrery (in part)."[33]

At Dartmouth, the orrery has been traditionally attributed to George Adams, instrument maker to King George III and one of the premier makers in London in the latter part of the eighteenth century. However, we recently discovered the mark *BWC* on the bottom of the brass disc, which may be an attribution. In the thriving London instrument scene of the late eighteenth century, instruments were often assembled from components made by different makers. Perhaps several shops supplied parts for this orrery.[34]

Although perfectly functional, this orrery is of a plain construction and most likely was manufactured as a cheap instrument for instructional purposes. Indeed, even Adams supplied "cheaper" orreries. The octagonal wooden base is worn and has some missing trim, as seen on the photograph of the right top surface. These time-related defects notwithstanding, the joinery on the base, although sound, is not of the highest quality.

The brass top is nicely finished on its upper surface, which is heavily patinated in a dull black finish. The lower unpatinated surface remains rough, as seen in the upper part of the next figure. The top has the usual divisions into zodiacal signs, but also has engraved on it a number of stars. They do not appear to represent any known constellations; they are of different sizes, suggesting the representation of magnitude.

The gear mechanism on the underside is supported between the top plate and another brass plate below it. The gears are of a "square-tooth" pattern. The gears and cam designed to simulate the motion of the

earth-moon system are fully visible on the top of the orrery (above). The outer planets are not gear-driven, but rather swing freely on arms rotating about the central, driven axis.

This orrery shows only the innermost six planets. In 1797, Union College in New York State received an orrery from W. & S. Jones that included Uranus (discovered by William Herschel in 1781).[35]

According to definitions offered by historian John Millburn, Dartmouth's is a "hybrid orrery" because it combines a planetarium (i.e., a "device representing the relative motions of the planets around the Sun") and an orrery (i.e., "a device representing the motions of the Earth/Moon system round the sun in three dimensions, incorporating the inclination of the Moon's orbit"). The terminology used to describe astronomical models since the eighteenth century has often been conflicting and confusing. A visitor to Dartmouth in 1787 remarked on the "planetarium" he had seen.[36]

Precession of equinoxes demonstration

EASTERN SCIENCE SUPPLY COMPANY
Boston, Massachusetts
About 1932
Height: 52 cm

2002.1.34771

Listed in ESSCO's 1932 catalogue for $36, this 8-inch terrestrial globe is mounted as a gyroscope. A movable internal counterweight can be adjusted to vary the period of precession as the globe spins on its axis. According to the catalogue, the device "shows principle and operation of a gyroscope, precession of the equinoxes, revolution of celestial pole about pole of ecliptic, nutation, as well as various phenomena of astronomy and physics."[37] The gores bear the imprint of *W. & A.K. Johnston, Limited*, a well-known Edinburgh publisher of cartographic material.

UNIVERSAL SUNDIAL

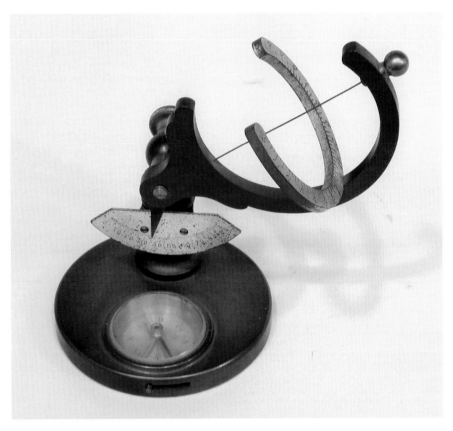

EASTERN SCIENCE SUPPLY COMPANY
Boston, Massachusetts
About 1932
Height: 18 cm

2002.1.34765

The universal sundial can be adjusted for latitude, making the wire gnomon parallel to the earth's axis. The magnetic compass (*MADE IN FRANCE*) permits aligning the polar axis of the sundial with the local meridian. The shadow of the gnomon wire is projected onto the arc, which shows the local solar time.

The illustration at the right shows the 1932 ESSCO catalogue description of this instrument, which listed for $12.

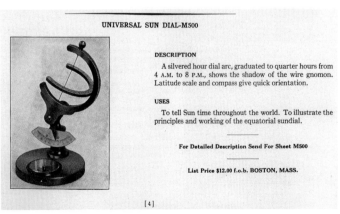

UNIVERSAL SUN DIAL-M500

DESCRIPTION

A silvered hour dial arc, graduated to quarter hours from 4 A.M. to 8 P.M., shows the shadow of the wire gnomon. Latitude scale and compass give quick orientation.

USES

To tell Sun time throughout the world. To illustrate the principles and working of the equatorial sundial.

For Detailed Description Send For Sheet M500

List Price $12.00 f.o.b. BOSTON, MASS.

[4]

ESSCO Rapid Reference Pictorial Bulletin, 1932

EASTERN SCIENCE SUPPLY COMPANY

ASTRONOMICAL AND SPECIAL SCIENTIFIC
EQUIPMENT

SALES OFFICE AND FACTORY, 7-9 HIGH ST., BROOKLINE, MASS.

POST OFFICE BOX 1414
BOSTON, MASS., U. S. A.

October 27, 1937.

Professor Richard H. Goddard
Dartmouth College
Shattuck Observatory
Hanover, New Hampshire.

Dear Sir:

Replying to your favor of the 22nd, we would advise that
we have shipped your order in full and included the sheets requested. We hope the shipment will reach you promptly and in good order.
Thank you for the courtesy.

The celestial eyepieces for the JT Telescope are $6.00
each. This price is higher than seems comparable, but is so because
we have not yet been able to purchase them on the quantity basis that
obtains for the other parts of the instrument. The telescope, without tripod, is $24.00 net, and the four celestial eyepieces would
be $24.00. This makes $48.00 for the total. Included in this is the
terrestrial eyepiece, which is part of the telescope assembly. If
this were not desired, it could be exchanged for a celestial eyepiece,
bringing the price to $42.00 for the telescope and celestial eye-
pieces.

These instruments show the rings of Saturn, moons of Jupiter
and double stars excellently well and have made a very lasting impression for their quality.

Re the camera. We built a complete mount and camera for
the Nanking (China) Observatory a while ago and made up four of the
cameras at that time. They were for a Ross-Lundin lens of 30" f.l.
and covered an 8"x10" plate. We are sending a print of the entire
mount which shows the camera clearly. We are enclosing a return
addressed label for your convenience, and would ask that you return
this photograph after you have examined it.

The camera is made of aluminum and is built as well as
we could design it for strength, rigidity and lightness. It has no
guiding telescope.

We sold one for $125.00 and it is in use at Mr. Gustavus
Cook's Observatory at Wynnewood, Pa.

You have a good shop there, and if you felt that you could
cut the lens end of the camera back to the 18" focal length, we would
supply the camera for $50.00, as we do not plan to do more work of this
type, and want to clear our stock. We could send the camera up for

Professor Richard H. Goddard-   2.

inspection, if you wished, at your expense, or you could see it here.

Also, we could cut the camera box to your specifications, and
make the adapters for the plate holder on a labor and material basis.
Our guess is that this would cost about $50.00.

We have the patterns for the guiding telescope frames and
the brackets attaching the camera to the mount. You could borrow
these, buy them, or we could have castings made as you wished.

In fact, all castings shown of the mount and camera were
made from patterns and coreboxes that we have, and we could loan
you what you might wish.

With the foregoing as a basis, we would be glad to have
your ideas.

Faithfully yours,

EASTERN SCIENCE SUPPLY COMPANY

By _H W Geromanos_
General Manager.

G/B.

There emerges evidence of an interesting relationship between the general
manager of ESSCO, H.W. Geromanos,
and Dartmouth professor of astronomy
Richard Goddard. Although we do not
have the beginning of this correspondence, evidently Goddard had inquired
about an ESSCO astrograph, a custom
instrument made for Nanking, China. An
astrograph is an equatorially mounted
camera designed specifically to photograph large areas of the night sky.
ESSCO had three more of them, and
wanted to get them out of their inventory. They were very willing to adapt the
cameras to Goddard's specifications.
Most surprising is the offer to lend
Goddard the patterns for the castings.
The complete mounted astrograph (below) was a substantial instrument.
ESSCO must have had a large amount
of money invested in the patterns, yet
was willing to lend them to Goddard.
The photograph below is a detail from
an original sent to Goddard by ESSCO.

ESSCO was known primarily as a supplier of simple pedagogical apparatus and
books. Yet this exchange indicates that

Photograph from H. W. Geromanos showing
astrograph

CORNELL UNIVERSITY
ROCKEFELLER HALL

DEPARTMENT OF ASTRONOMY

ITHACA, N. Y. May 3, 1940.

Prof. Norman E. Gilbert,
Dartmouth College,
Hanover, N.H.

Dear Professor Gilbert:-

        You are undoubtedly aware that the Eastern
Science Supply Company will be unable to continue as a
commercial concern, because the expense of conducting the
business has in general exceeded the income received from
the sale of the apparatus and supplies.

        I think all of us who teach courses in elementary
astronomy realize that it will be difficult or impossible
to get the supplies needed, if Mr. Geromanos sells his forms,
models and special machinery developed to produce what many
of us have found so essential as teaching aids.

        During the past year, I think Mr. Geromanos' own
correspondence has made it clear that he cannot continue the
business on his own resources any longer.

        He has as near assurance as can be given, before
actual application for a grant is made, that if a responsible
Board of Directors applies for a grant from the Carnegie
Foundation for a fund to carry on the production of the well-
established laboratory supplies, which Eastern Science Supply
Company has produced, such a charitable organization would
be supported on a modest scale to meet the needs of teachers
of elementary astronomy which cannot be met by a commercial
organization.

        The plan is to get the necessary number of Massa-
chusetts citizens, who know the Eastern Science Supply Company, to associate and ask for a charter for the new company.
These people would be the nucleus of a Board of Directors,
i.e. the Executive Committee of the Board, other members of
the Board to be selected from teachers of astronomy all over
the country. Such members would not be required to attend
Board meetings but could give advice when requested by the
Executive Committee or on their own initiative.

        The members of the Board of Directors, of course,
have no financial responsibility and only the small Executive
Committee would be responsible for the selection of the
Manager. The rest of the Directors would merely act in an
advisory capacity to the Executive Committee and through
them to the Manager.

        I write to ask if you will consent to act as such
a member-at-large of the Board of Directors, if the Executive Committee, when and if formed, should select you.

        If enough teachers of astronomy all over the country
evince enough interest in the formation of the new company
to show the Carnegie Corporation that the work is worth
their consideration, then the Executive Committee will apply
for a grant, which, if made, will make possible the continuation of the fine service which Mr. Geromanos has rendered
to teachers of astronomy for the last two decades.

        If this cannot be done, we shall be left where we
were before 1920. The situation will be worse, in fact,
than it was before 1920 because there are so many more of us.

                        Very sincerely yours,

                        S. L. Boothroyd

                        S. L. Boothroyd.

in the 1930s they also had the capacity to make large precision equipment for astronomical research.

When ESSCO was threatened with financial failure in the late 1930s, a group of American astronomers tried to obtain support from the Carnegie Foundation to maintain the company so that it could continue to manufacture their products. In June of 1940, Richard Goddard replied to this letter, indicating his strong support for the effort.[38]

The Eastern Science Supply Company was bought in 1948 and changed its name to Eastern Science Company, in which form it survives to the present.

49

MACHINE SHOP

Interior of Wilder machine shop, 1947

In his 1937 letter to Richard Goddard, H.W. Geromanos refers to Dartmouth's machine shop as a "good shop." The photograph shows Walter Durrschmidt, the department's machinist, at work in this shop, located in Wilder Laboratory, home of the Department of Physics and Astronomy. Durrschmidt worked for the department from 1930 to 1955. The clock tower visible through the window is that of Baker Library.

Equatorial star finder

C.J. KULLMER
Syracuse, New York
1911
Height: 29 cm
Model 3

2002.1.34777

Charles Julius Kullmer, a professor of German at Syracuse University since 1905 and long-time lecturer on popular astronomy, invented this device which he allegedly sold by the hundreds to American universities, colleges and schools, as well as to prominent business and civic leaders. An undated testimonial lists among its purchasers no less than Admiral R. E. Peary, William Waldorf Astor, several bishops and mayors, one U.S. senator, bankers, the president of the Great Northern Railway, and the president of the Royal Astronomical Society of England.

Catching the wave of popular interest following the return of Halley's Comet in 1910, Kullmer hyped his invention in cosmic terms: "Four hundred years ago Copernikus discovered the explanation of the seemingly complicated movements seen on the heavens; nowadays everybody knows this explanation, but very few know what the explanation explains. It is the duty of every intellectually inclined person to make clear to himself his real relation to the rest of the universe. Now, honestly, when you look up at the star-lit heavens, what do you really know about it all? Wouldn't it be worth to you the price of this little instrument to have all those difficulties cleared away? The price is only $5 express paid." *Scientific American* praised the simplicity of the device; one purchaser lauded it as "a really great contribution to human happiness." The name *COPERNICUS* is emblazoned across the shield placed above the signature on the base of the Model 3.

Like the medieval torquetum, the plane of the central disk is set parallel to the celestial equator. To locate a constellation or star at a particular time, the metal slide is set to the day in question and the movable celluloid disk is rotated to the hour in question. The pivoting indicator is then rotated to the constellation desired (right ascension) and the pointer is set to the angle for that constellation as indicated on the celluloid disk (declination). The constellation will then be visible by looking directly along the pointer. On our instrument, the calibrations on the pointer are absent; it cannot easily be set to a specified declination.

Kullmer's thirty-page instruction manual, *Star Maps and Star Facts*, introduced users to basic concepts of spherical trigonometry and the motions of the celestial sphere. Instructions for finding the positions of the bright planets from 1910 through 1925 were also included.

By 1911, the Göttingen maker, F. Sartorius, was advertising a similar Sternfinder (star finder) for "teaching and amateur astronomers" that had been especially approved by the Prussian Educational Ministry.[39]

ACHROMATIC REFRACTING TELESCOPE

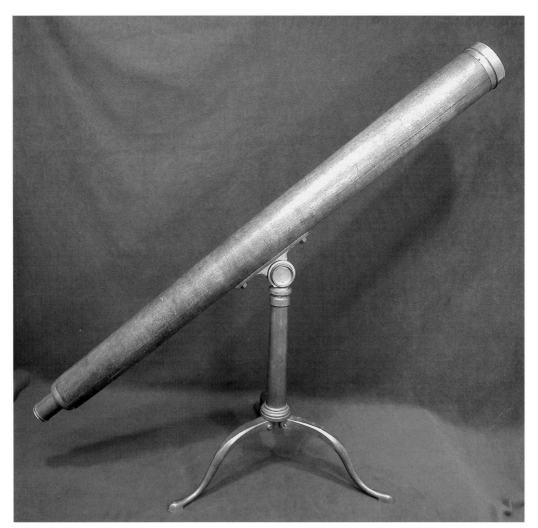

W. & S. JONES
London, England
1792-1800
Diameter of objective: 3 inches; Length: 42 inches
Exhibited in "Illuminating Instruments," Hood Museum of Art,
Dartmouth College, 2004.   2002.1.35202

W. & S. Jones' 1794 catalogue offered a "new-improved 2 1/2 feet achromatic refractor, on a brass stand, mahogany tube, with two sets of eye-glasses, one magnifying about forty times for terrestrial objects, and the other about seventy-five times for astronomical purposes, packed in a mahogany box. £9 9s 0d."[40] The model with a 3 1/2 foot mahogany tube seen above, listed for £16 16s. This telescope has one internal stop to reduce unwanted reflections from the inner surface of the tube. Mounted

A similar telescope and mount, 1856

54

on a pillar-and-claw stand, the legs fold so that the entire instrument fits into its case without disassembly. This design proved to be a successful table-top mounting for small refractors and was used well into the nineteenth century, as seen in the inset on the previous page.

The achromatic lens, which reduces chromatic aberration by utilizing glasses of different refractive indices, had been patented by John Dollond of London in 1758. In the late eighteenth century, a three-inch lens such as found on this telescope was of significant size. This Jones telescope was very popular; in 1796, Union College ordered the same model.[41]

When in 1869 Charles A. Young traveled to Burlington, Iowa, to observe the total solar eclipse, he used the College's 4-inch Merz telescope for his spectrometer, but also brought along the Jones achromatic refractor.

A detail of the photograph of the Iowa expedition shows the latter telescope standing prominently on a fancy wooden table in front of crudely constructed observation huts. Young outfitted the telescope with a screen, supported by a light wire framework at the eyepiece end. The objective is surrounded by an opaque square to cast shadow on the rear screen. The Sun's image

1869 eclipse expedition; photograph detail shows achromatic telescope fitted with projection screen. Diagram at right, from Charles A. Young, *The Sun*, 1881, illustrates this setup.

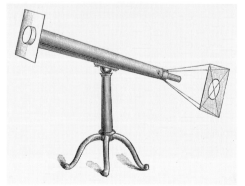

is projected, permitting several observers to examine it simultaneously. Since it requires no filters and obviates direct viewing which, even with filters, can damage the retina, the technique is still popular. Widely used to demonstrate sunspots, here it would have been used to show the progression of the eclipse.

In the above photograph, the woman at the right is Maria Mitchell, a faculty member at Vassar College and pioneering woman astronomer who also had journeyed to Iowa to observe the eclipse.

56

SHATTUCK OBSERVATORY

Shattuck Observatory, about 1905

Designed by Ammi Young, Dartmouth, 1841 (honorary)
1853-1854
Rotunda: 20 feet outer diameter; north and south wings: 20 x 16 feet; east wing: 35 x 16 feet; dome: 18 feet inner diameter

The Shattuck Observatory represents Dartmouth's first major investment in science as a complex infrastructure of instruments, books and buildings that would support teaching, research and outreach to the local community. In creating the observatory, Dartmouth followed national and international trends and established a space in which its first internationally acclaimed scientific research would be conducted.

Reviewing in 1832 the "progress of astronomy," the Cambridge astronomer and soon-to-be Astronomer Royal George Airy noted: "I am not aware that there is any public observatory in America, though there are some able observers." Several years earlier, President John Quincy Adams, urging the U.S. Congress to create a national astronomical observatory, had made the same point: "It is with no feeling of pride, as an American, that the remark may be made that, on the comparatively small territorial surface of Europe, there are existing upward of one hundred and thirty of these lighthouses of the skies; while throughout

Shattuck Observatory in 2003

Lens from original refractor by G. Merz & Söhne
Unsigned (Joseph Fraunhofer?)
Munich, Germany
Before 1825
Diameter: 6.4 inches   2002.1.35417

the whole American hemisphere there is not one." Rejecting the idea that science was a federal responsibility, Congress ridiculed Adam's suggestion. "Lighthouse in the sky" became a popular term of derision for harebrained, impractical ideas.

Despite Congress's assessment, however, an "observatory movement" swept antebellum America. Between 1836 and 1856, many entrepreneurial astronomers and their patrons erected more than twenty observatories, specially designated, newly constructed buildings with permanently mounted equatorial and transit telescopes. Local boosterism, popular enthusiasm for astronomy, practical need for accurate time-keeping, and an emerging ethos of internationalism and research all drove this movement. In an 1856 survey of American astronomy, New York University astronomer Elias Loomis especially emphasized the latter: "College

58

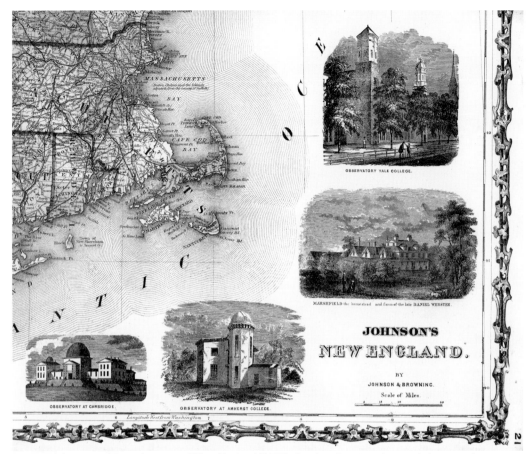

Johnson and Browning map of New England (1862) showing vignettes of several observatories

professors ... are in danger of settling down into mere retailers of other men's ideas without aspiring to add anything to the stock of human knowledge, unless they are surrounded by institutions whose leading object is the increase of knowledge. An astronomical observatory, therefore, is a centre of genial influence, which directly or indirectly imparts life and efficiency to all the subordinate institutions of education." In addition to twelve college observatories, the movement also yielded several private and corporate observatories and the U.S. Naval Observatory in Washington, D.C.[42]

Ira Young, professor of mathematics and natural philosophy from 1833 until his death in 1858, pushed Dartmouth into this movement. In 1841, Dartmouth's trustees agreed to raise $30,000 by subscription, to establish two new professorships, add to the philosophical apparatus and library, and erect an astronomical observatory. As those funds slowly accumulated, Young in 1846 petitioned the trustees for $4200 to equip a "small astronomical observatory" with an equatorial telescope, transit circle, comet seeker (a small telescope with a relatively wide angle of view) and astronomical clock. He listed other colleges—Amherst, Wesleyan, Williams, Princeton, Western Reserve College, and even the Philadelphia High School—that had provided from six to ten thousand dollars for observatories. Apparently persuaded by this rhetoric of competition, the trustees authorized $2300 to enable Young to visit other observatories and to purchase apparatus. After corresponding with the Philadelphia astronomers who in 1840 had received the first refractor in America made by Merz & Mahler, Young purchased a 6.4-inch telescope of focal length of 8 feet 7 inches and an astronomical clock from this Munich firm, then known as G. Merz & Söhne. The apparatus arrived in 1848 and, after Congress agreed to forego the import tax of $700, was mounted by 1849 in a temporary 28 x 13-foot building with sliding roof erected in the garden of Young's home.

The objective lens (see illustration) "was originally ground and polished by [Joseph] Fraunhofer," according to a 1944 letter from Walter S. Adams, then director of the Mount Wilson Observatory in California.

The original Merz telescope and mount, about 1870

Georg Merz had apprenticed with Fraunhofer for eighteen years before assuming responsibility for the firm's glassmaking at Fraunhofer's death in 1826; in 1839 Merz purchased the firm. The two-element achromatic lens is air-spaced. The best relative angular position of each element was often established empirically by rotating one element with respect to the other until maximum resolution was achieved, and marking these positions with registration marks on the edges. In June 1969, Allen King disassembled the

Merz lens and noted: "Found markings on edges—one place has date of 17 Janu. 1878. Word 'tube' appears near outside edge of negative lens—'10 feni' on edge of both disks at one place."

With the telescope Merz provided eighteen eyepieces, magnifying from 36 to 940 times, and two micrometers for measuring apparent distances between stars. The equatorial telescope mount also came from Merz and is of the "German" type invented by Fraunhofer (see illustrations). The mount has an hour circle 9.5 inches in diameter and a declination circle 13 inches in diameter. Declination could be read to 10 arc secs, right ascension to 4 time secs. A weight-driven clock moved the telescope.

The College Catalogue for 1849-50 proudly announced the new instrument: "The lectures in astronomy are accompanied by celestial observations and instructions in the use of instruments. The splendid telescope obtained during the past year, ranking as the third in the United States, in magnitude and power, supplies

Astronomical clock face
Utzschneider und Fraunhofer
Munich, Germany
1848
Diameter: 30 cm

2002.1.35455

61

Fraunhofer mount detail

important facilities for these purposes." By Young's reckoning, the Dartmouth lens was surpassed only by the Merz telescopes at the new observatories in Cambridge (15 inch) and Cincinnati (9.5-inch). People flocked to Young's garden to peer through the new telescope. "The instruments have been in almost constant use, when the weather would permit, either for the instruction of the students, the gratification of visitors, or for regular observations," reported Young in 1850.[43]

Desiring a larger structure and better site for the observatory, Young in 1852 approached the prominent Boston physician, George C. Shattuck, Dartmouth 1804, for help. Having founded the *New England Medical Journal* and donated $26,000 to Harvard's Medical School, Shattuck was known for his charitable instincts. He agreed to give Dartmouth $7000 toward the construction and equipping of an observatory and $1790 for an astronomical library, provided that tthe trustees contribute another $4000, send Young to Europe to purchase the requisite apparatus and books, and commission Young's brother, Ammi, to design the building. The trustees accepted these conditions. In 1853, Young and his son, Charles, made the European purchasing trip, and in 1854 the Shattuck Observatory became operational.

Although widely known for his Greek-revival style, Ammi Young, who had designed several earlier Dartmouth buildings, Boston's Custom House, and as architect of the U.S. Treasury Department many other federal buildings, planned a simple, unornamented brick structure for his brother. Situated on a knoll seventy feet above the campus, the building (still standing--see illustration) included a two-story, domed rotunda with the equatorial telescope on the upper and a library on the lower level, a meridian transit room to the east, a prime vertical transit room to the north, and a bedroom and additional observer's room with a slit roof to the south (later to become the telegraph room). Revolving on six cannon balls, the 2800-pound dome allegedly could be turned with a force of only six pounds. The building's layout, with a two-story rotunda protruding from the side of a long, rectangular structure, is unique among the buildings erected during America's "observatory movement." Extant notes written by Charles and Ira Young indicate that they explicitly copied features of a dome designed by William Lassell, the English brewer and amateur astronomer, for his private observatory in Liverpool. During the 1853 European trip, the Youngs visited Lassell's observatory (as well as observatories in Edinburgh, Oxford, Paris, Brussels, Leipzig, Berlin and Munich).[44]

On that trip, the Youngs further equipped the observatory by purchasing a 30-inch meridian circle with 4-inch transit telescope from the London makers, Troughton & Simms, and a 4-inch comet-seeker, also by Merz, that was mounted equatorially. They also bought books for the library, another integral component of the observatory. By 1857, it contained roughly 500 volumes on theoretical and practical astronomy, many periodicals from European observatories, and classic texts in the history of astronomy. Like the Merz telescope, these books represented Dartmouth's first major acquisition of tools intended for scientific research rather than teaching.

Equatorial telescope mount
Unsigned (G. Merz & Söhne)
Munich, Germany
1848
Height: 36 inches

2002.1.35452

Fraunhofer's 1825 mount for the
Dorpat telescope

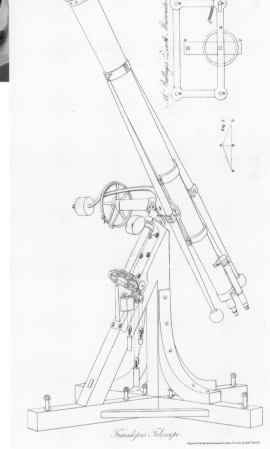

The observatory building cost $4800 to erect. Young paid $2200 for the Merz telescope and astronomical clock, $1400 for the meridian circle and transit, $175 for the comet seeker, and $940 for books. Adding freight and Young's travel expenses, the Shattuck Observatory required a total outlay of about $10,000. In 1846, Young had estimated that Dartmouth since its origin had spent a total of only $2300 for philosophical apparatus. The Shattuck Observatory clearly opened a new era in the College's involvement in science. Although he himself would never publish any original research, Ira Young created the tools that would enable his son, Charles, to become Dartmouth's first world-renowned scientist during his tenure as professor of astronomy from 1865 to 1877.[45]

Merz had been the principal supplier of refractors for the American "observatory movement." Yet by the 1850s, astronomers began to criticize the quality of the Munich lenses. Complaining that the Merz telescope was too small and unsteadily driven to obtain satisfactory photographic results with his new Clark telespectroscope (see entry, page 83), Charles Young in 1871 replaced his father's telescope with a 9.4-inch refractor of 12 feet focal length and a new clock drive, both made by Alvan Clark & Sons. The Clark lens, Young later wrote, "is considered by the makers a little under-corrected, but suits my own eye very well, in pleasant contrast to the violent over-correction of a 6-inch Munich glass which I had used for several years previous. The spherical aberration is very perfectly corrected. The curves are essentially those of Littrow...a nearly equiconvex lens of crown, and a nearly plano-concave flint."

The tube of the new telescope, still mounted under the Shattuck dome, consists of steel plates riveted into two cone-shaped elements, with the widest parts at the saddle where the tube is attached to the declination axis, tapering to the objective and eyepiece ends. Such a double cone provides exceptional rigidity for its weight. Although not an original design (having been used already in the seventeenth century), the Dartmouth telescope is one of the earliest Clark tubes to use this construction. The Merz comet seeker was (and remains) attached as a finder to the Clark tube, and the old Merz equatorial mount was jury-rigged to support the new telescope.

Clark & Sons charged Young $4000 for the new apparatus but accepted Dartmouth's old Merz refractor as partial payment, reducing the cost to $2500, money Young raised by private subscription. The 6.4-inch Merz lens found its way into other telescopes, including the one used by É. Léopold Trouvelot, a skilled observer and astronomical illustrator who had a private observatory in his Cambridge (Massachusetts) garden, worked closely with the Harvard astronomers and the Clarks, and in 1878 joined a U.S. Naval Observatory party to observe the total solar eclipse with his "refracting telescope, by Merz, of six and one-third inches aperture." Eventually the lens came into possession of the Theosophical University in California, whose trustees in 1969 returned the lens to Dartmouth as a bicentennial gift for the College.[46]

In 1904, John M. Poor, professor of astronomy and director of the Shattuck Observatory from 1898-1933, installed a second transit telescope of 3-inch aperture, made by G. N. Saegmuller of Washington, D.C. In 1908, Poor adopted the Clark telescope for photography by engaging Carl A. R. Lundin of Alvan Clark & Sons to make a photographic corrector (see entry, page 71), provide a new equatorial mount and replace the weight-driven clock with an electric motor, all for a cost of $3600. The earlier brick equatorial pier was replaced by one of poured concrete, with recesses for the electric drive mechanism.

By this time, however, the 9.4-inch Clark had reached the end of its life as a tool for original research. Despite his renovation of the Clark refractor, Poor would publish no astronomical papers based on Shattuck

Bly's albumen print stereo view of the original Merz refractor

64

The recto and verso of Bly's stereo card showing the Merz refractor and a list of Bly's views

observations. The last such published work appears to have been Edwin B. Frost's observations of comet a1890. Both Poor and his successor, Richard H. Goddard, increasingly emphasized meteorological work. Goddard, who retired in 1963, was the last director of Shattuck Observatory. Thus, after an active "research life" of about twenty years from 1870 to 1890, the observatory and its century-old instruments subsequently have been used primarily for teaching and public viewing. In 1959, the original dome was replaced by the present dome of galvanized steel, which is rotated electrically. Other efforts to expand or re-equip the observatory have failed.

During its initial decades when much of New Hampshire remained open farmland, the Shattuck Observatory overlooked the town of Hanover and stood as a highly visible symbol of science and learning. Around 1870, the Hanover photographer H.O. Bly produced a set of stereo view albumen prints featuring places of local interest, including the observatory. One shows the 6.4-inch Merz refractor with Young's telespectroscope attached. Another view of the refractor, framed by the doorway, reveals the ribs of the original wooden dome. A third illustrates the transit room with the Troughton and Simms instrument mounted on massive granite pedestals. The original observing couch, covered with velvet, appears between the pedestals.

9.4-inch Clark refractor on mount installed in 1908, photographed in 2004

Over the past century as farming declined in New England, forests have returned across the region. In Hanover, trees have encroached upon the observatory (see page 58), rendering it largely invisible from the town and restricting its viewing to objects at higher altitudes. Similar situations at other old observatories prompted Allyn J. Thompson, a major figure in postwar amateur telescope making, to write:

> But neighbor's trees grew up and o'er it,
> The stars in vain their light would send
> To seek their fickle faithless friend.

In their oft-reprinted popular book on telescopes, George Z. Dimitroff and James G. Baker wrote: "A telescope is more than a machine; it is an enterprise around which clusters the life of an institution." Aptly supporting this claim, the Shattuck Observatory over the past century and a half has served as a site for more than astronomy and meteorology. From the 1860s through the 1920s, students often lived in the

The original Merz "comet-seeker" is still attached to the Clark refractor. Note the tapered, riveted Clark tube.

building, working as custodians or weather observers (see page 70). In 1865, one of the earliest boarders reported to a friend:

> I am now rooming at the Observatory. There is only one room in this building occupied by students & this is given out free to the best student in the Senior Class. The Sen. wanted to be absent three months & asked me to take his place which I accordingly did. I have the keys to every room in the building and have complete control of all the instruments. There are two telescopes -- both made in Germany. The best one cost $3000, & is a very fine instrument. There is also a very fine Transit which cost $7000. The Siderial Clock is regulated by this. There are many more instruments but I have no time to mention them. If you will make me a call I will let you look thro' the Spy Glass all night if you want to.

Edwin B. Frost, later astrophysicist at the University of Chicago and director of the Yerkes Observatory, fondly recalled his teaching sessions in Shattuck during the 1890s:

> The men seemed to enjoy the appointments even although they were sometimes scheduled to meet at the observatory as late as 2 a.m. At that hour the members of the class apparently aroused every dormitory, and the noise of the advancing column over the temporary board walks with the thumping of the canes which the junior class affected at that time of the year could be heard through the whole village. Then for an hour or two the class of eighty or more men would take their turns in looking through the telescope in close marching order, and when the exercises were over the return to the different dormitories was made without regard to those who would fain sleep.

And during Director Poor's tenure, small groups of faculty, students and town friends would gather at the observatory on Sunday evenings for oyster stew and strong coffee. The observatory on the hill provided a place for both science and sociability.[47]

Bly's stereo view of the transit instrument. This instrument remains essentially unchanged, on the original granite piers. The clock in the corner is still in place and is the one adapted to send signals to the recording chronograph (see page 72).

ANEMOMETER AND RECORDER

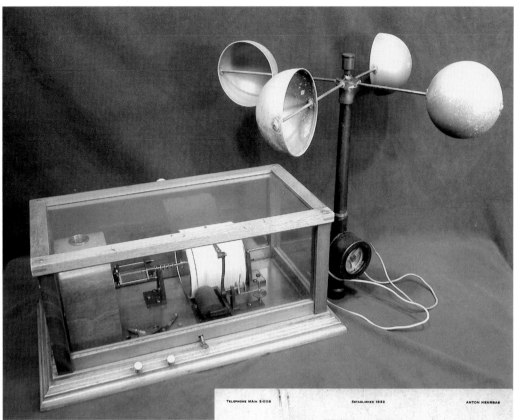

HENRY J. GREEN
Brooklyn, New York
About 1930
19.5 x 44 x 25 cm
Serial numbers: 177 (recorder),
372 (cups)   2002.1.35214

These instruments were supplied by Henry J. Green, an American instrument-making firm founded in 1832. The anemometer is Green's Model 323a; it is connected to the recorder Model 324a, which keeps a daily record. This device only records wind velocity, not direction. Special ink is applied directly on a paper affixed to a drum driven by a clockwork mechanism. Additional supplies of ink, as well as lots of paper sufficient for a year, were sold separately.

TELEPHONE MAIn 2-0116   ESTABLISHED 1832   ANTON NEHRBAS

HENRY J. GREEN
SUCCESSOR TO JAMES GREEN
MANUFACTURER OF
METEOROLOGICAL INSTRUMENTS
TO THE
U. S. GOVERNMENT DEPARTMENTS, UNIVERSITIES, ETC.
1191 BEDFORD AVENUE

BROOKLYN, N. Y., U. S. A.

June 15, 1934

Richard H. Goddard, Dept. of Astronomy
Shattuck Observatory -- Dartmouth College
Hanover, New Hampshire

Dear Sir:

Regarding the list of instruments you desire directions for instalation, desire to say that the #609 double register is for anemometer #323a and wind vane #324b--#323b anemometer cannot be used with the register since it makes contact for every 1/10th of mile of wind velocity, whereas #323a makes contact for every mile.

The #324 anemometer register cannot be used with the #593 tipping bucket rain gauge but with #323a Anemometer only.

A register #593e for the tipping bucket rain gauge will cost $163.00.

Wind vane #334a is an interior indicating direction indicator, hand being attached to vane rod and dial attached to ceiling underneath support of vane. This vane has no electric contact.

Any other information with reference to the above and not contained in blue prints or directions herewith will be gladly given upon request.

Very truly yours,

Henry J. Green
H.

Enclosures:

Models 323a and 324a remained unchanged in Green catalogues from the 1910s through the 1950s.[48]

The anemometer cups are aluminum and still spin freely. A spring-driven mechanism, signed by the American clock manufacturer Seth Thomas, turns the chart by means of a worm screw.

Dartmouth has maintained a weather station, at the site of the Shattuck Observatory, since 1853, producing the oldest continuous weather record in the United States. Weather records, consisting of observations of temperature, rainfall, snow, wind, sunshine and clouds, were originally maintained by students who, in return for this work, were given lodgings in the observatory. This photograph shows (seated) Frank R. Sanborn (Dartmouth College, 1887), and (in hammock) Charles F. Chase (Dartmouth College, 1885). From 1887 to 1889, they were students at the Thayer School of Engineering and acted not only as weather recorders but also as custodians of the observatory, working under the supervision of Edwin B. Frost (Dartmouth, 1886), who would serve as professor of astronomy and director of the observatory from 1892-98.[49]

The Shattuck Observatory had living quarters for the weather recorders.

70

PHOTOGRAPHIC CORRECTING LENS

Unsigned (Alvan Clark & Sons)
Cambridgeport, Massachusetts
1909
Diameter: 10 inches   2002.1.35392

A two-element achromatic lens (typically crown and flint glass) brings into focus only a few colors of the spectrum. The other colors are strung out along the optical axis like beads on a string. A telescope lens designed for visual work generally is "figured" or shaped so that yellows and greens, colors to which the human eye is most sensitive, are focused. However, early photographic plates were most sensitive to blue light and thus had to be placed along the optical axis where the blue part of the spectrum was in focus. A photographic corrector such as this one was designed to replace one of the elements of the objective, thereby providing a means of modifying a refracting telescope from visual to photographic use.

The objective lens in the Shattuck Observatory's 9.5-inch Clark objective refractor is corrected for visual use. In 1908, John Merrill Poor, assistant professor of astronomy at Dartmouth, "decided to adapt the equatorial to photography." As Poor later explained, "this involved addition to the optical parts, an entirely new mounting and clockwork, and some changes in the building and pier. Mr. C.A.R. Lundin, head of the Alvan Clark & Sons Corporation, designed and made the photographic 'corrector' which consists of a 9.4-inch disc of flint glass, which when used in place of the original 'visual' flint disc, transforms the telescope into a photographic instrument of 10 1/2 feet focal length. The focal length of the instrument used as a visual is 12 feet. The cost of this change in the optical parts was $600. ($550 for the flint disc and fifty dollars for a new cell)."[50]

Lundin documented the corrector as follows (the revolution numbers refer to spherometer [see entry, page 114] readings):
Dec 1908
New flint lens for a photographic corrector for the Dartmouth [sic] telescope, Prof Merril Poor
Flint No. 5918
Old crown curve   +1 rev 92 1/2 div.
" flint   "   inside -1 rev 99.6 div
"   "   "   outside -10.4 div
Focus visual 143" from back of flint
New flint   "   inside -1 rev 99.6 div
"   "   "   outside + 5. div
Focus using corrector in place of old flint 126 13/16" from flange.
Thickness of new flint on edge 11/16"
Price for corrector $550.00 Finished Feb. 2, 1909.[51]

By the early twentieth century, yellow-sensitive photographic plates were being developed, which obviated the need for blue correction.

Drum chronograph

Unsigned (Alvan Clark & Sons)
Cambridgeport, Massachusetts
1872
35 x 56 x 30 cm

2002.1.35453

At the 1851 Crystal Palace exhibition in London, a total of thirty-one Council Medals were awarded for "philosophical instruments." Only one of those went to an American exhibit, to William Bond and his son George of Cambridge, Massachusetts, for their "apparatus for observing transits by means of a galvanic current." By employing a seconds-pendulum and a mercury trough, they arranged a circuit that would create a one-second make-break circuit. This would activate an electromagnet, driving a pen that marked one-second intervals on a rotating drum. Using a telegraph key, an astronomer could also break the circuit and thereby mark the observed times of transits against the comparison of one-second ticks.

First developed at the U.S. Naval Observatory in 1849, the Bonds by 1851 had

72

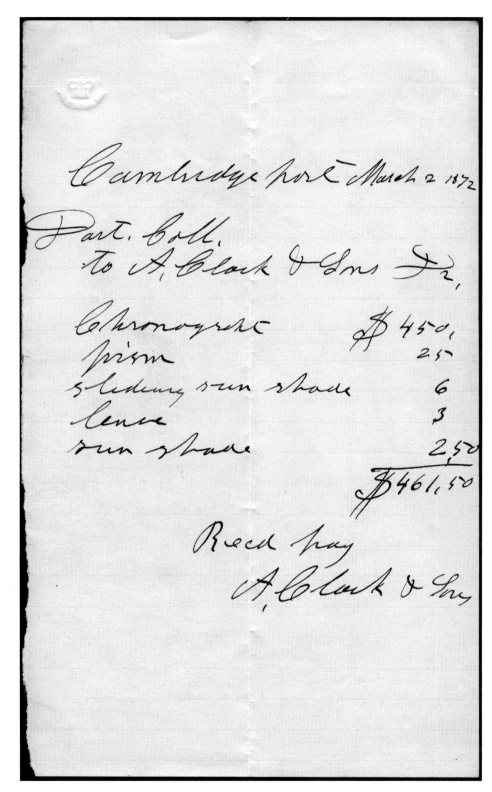

Cambridgeport March 2 1872

Part. Coll.
 To A, Clark & Sons Dr

Chronograph $ 450,
prism   25
Sliding sun shade   6
lense   3
sun shade   250
        $461,50

Recd pay
A, Clark & Sons

1872 receipt from Alvan Clark & Sons for chronograph

installed the chronograph at the Harvard College Observatory. Within several years the Greenwich Observatory followed suit. By measuring altitudes of a celestial object with a transit telescope, fixed in the plane of the meridian, and recording the time of passage, the position of the object can be computed. Transit measurements comprised a major portion of the work at most major nineteenth-century observatories; hence the Crystal Palace jury's enthusiasm for the Bonds' chronograph—it will "in all probability, form a new era in astronomical observation."[52]

A weight-driven clockwork, with its speed regulated by a centrifugal governor, rotates the drum and the screw, which advances the pen to generate helical traces on the paper. In the Shattuck Observatory, where these instruments currently reside, the transit clock made by local craftsmen has a temperature-compensating mercury-weighted pendulum. The clock hangs on a wall several meters from the chronograph.

The design of this chronograph apparently was very successful. An instrument appearing very similar to this exemplar was used in 1888 when the U.S. Coast and Geodetic Survey measured

Astronomical clock, compensating mercury pendulum
Unsigned (Johnson Brothers)
Sanbornton, New Hampshire
1872
150 x 49 x 29 cm   2002.1.34545

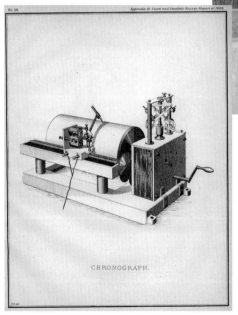

Similar chronograph used in 1888 by
U.S. Coast and Geodetic Survey

transits on the Hawaiian Islands. Interestingly, Charles A. Young's first published paper, appearing in 1866 only months after assuming his Dartmouth professorship, proposed an improved chronograph that would print numerical values of times to the nearest hundredth of a second. Apparently Young never managed to build a working model of this device, and instead ordered this chronograph from Clark & Sons, for which he paid $450.[53]

74

SUPERHETERODYNE RADIO RECEIVER

Lafayette
New York, New York
About 1935
53 x 41 x 31 cm

2002.1.35456

Hertzian waves, traveling at the speed of light, are ideal carriers of time signals. They enable chronometers to be synchronized because Greenwich Mean Time can be broadcast and every ship or aircraft with a receiver can obtain accurate time without elaborate instruments. The U.S. Naval Observatory by 1914 was broadcasting standard time signals from the Navy wireless telegraphy station, NAA, in Arlington, Virginia. Synchronized second pulses were broadcast for the five minutes preceding noon and 10:00 PM each day, E.S.T. According to a 1914 publication of the National Bureau of Standards, the average error in the Navy's signals was less than .05 seconds and the broadcast could be received for more than 1000 miles from Arlington.[54]

This superheterodyne receiver thus brought time information directly into Dartmouth's Shattuck Observatory, freeing astronomers from the need to maintain standard time by mechanical clocks, chronometers, telegraph signals or transit instruments.

Invented in 1918 by Edward Armstrong, then a member of the U.S. Army Signal Corps, superheterodyne circuitry converts incoming weak electromagnetic waves of high frequencies to a constant low frequency before detection and amplification. The technique greatly improves both reception and tuning and is employed by most of today's radio receivers.

75

Prisms

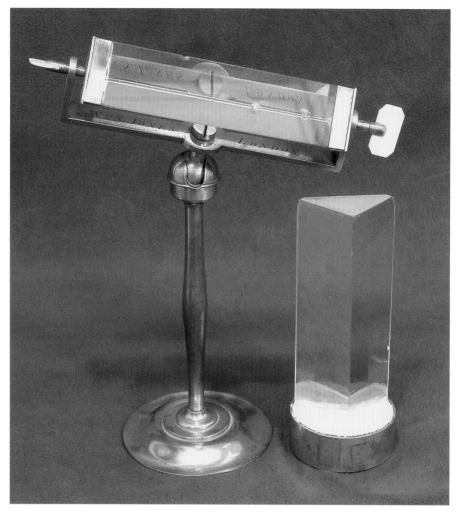

W. & S. Jones (left)
London, England
About 1795
17.5 x 14 cm   2002.1.35199

Unsigned (made for U.S. military)
About 1950
13 cm (without cell)
Both exhibited in "Illuminating Instruments," Hood Museum of Art, Dartmouth College, 2004.   2002.1.34564

Since 1672 when Newton published his controversial study of spectral colors, prisms became a common tool in experimental natural philosophy. By breaking white light into its colored constituents, prisms enabled fundamental research on light. By the mid-nineteenth century, studies of the spectrum had provided new insights into the chemical elements and the composition of stars and nebulae, thereby contributing to the rise of astrophysics.

The Jones prism, mounted on a brass stand with a ball joint, was made specifically for demonstration purposes. The other prism was manufactured for the U.S. military for use in a tank periscope. Two of

76

TANK PRISMS

THE BIGGEST OPTICAL VALUE WE HAVE EVER BEEN
ABLE TO OFFER

A War Surplus Item Priced To Sell Now!
BUY AT ONCE WHILE YOU CAN !

90-45-45° Prism - Finely ground and Polished

THE STORY - In order that the tank driver shall not get shot in the face, tanks are equipped with periscopes. Some of these are made with the mirrors described elsewhere in our listings and some are made with prisms. Normally these prisms would retail for about $30.00.

PERFECT PRISMS - Stock #3004 - Price $2.00 - Silvered Tank Prism
Stock #3005 - Price $2.00 - Plain Tank Prism

The following prisms are seconds for slight chips. No exchanges can be made for those who purchased tank prisms prior to this change.

Stock #3100 - Price $1.00 - Silvered Tank Prism
Stock #3101 - Price $1.00 - Plain Tank Prism

USES FOR TANK PRISMS

1. A very practical use these can be put to, is to turn one into a desk name plate. If you have a desk you shouldn't miss this opportunity! Make them up for your business friends for an unusual, fascinating gift item. Gold cut-out letters may be purchased at many stationery stores. They also make an excellent paper weight.

2. They are useful for experiments, class room demonstration at high schools, colleges, camera clubs, etc., and make an excellent gift item.

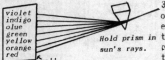

Hold prism in sun's rays.

3. If you so desire you can remove the silvering from the bottom of the prism to make it a plain prism. One of the most interesting experiments you can make with a plain prism is to use it to project the spectrum. White incident light as it passes through an unsilvered prism is broken up into a band of primary colors known as the spectrum. This is a beautiful sight! The drawing shows how this is done.

A WORLD OF COLOR - By looking through a plain tank prism at a certain angle you can see a world of all colors everywhere. An amazing sight!

FREE: *With each purchase of stock No. 3004 - a copy of our booklet on prisms which describes different kinds and tells how to use tank prisms to make periscopes, desk name plates, and how to saw them to smaller sizes.*

Undated ephemera (probably just post-WWII)

these right-angle prisms, each mounted in a heavy metal frame, enabled the tank crew to look outside without exposing themselves. For a right-angle prism, light entering one side will strike the hypotenuse, be totally internally reflected, and exit the other side. Such a prism deflects the path of light by ninety degrees. Vast quantities of these prisms were made during WWII and appeared on the surplus market after the war. A recent catalogue (2004) still offers these WWII tank prisms for sale (below).

WW II US ARMY TANK PRISM

These were spares to replace damaged prisms in tank periscopes. Prisms are mounted and in great condition. 6-3/8" long by 2" high by 1-1/2" wide at the base. Great collectable, or put two together to actually make a quality periscope.
L1814 $6.50

CARBON BISULFIDE PRISMS

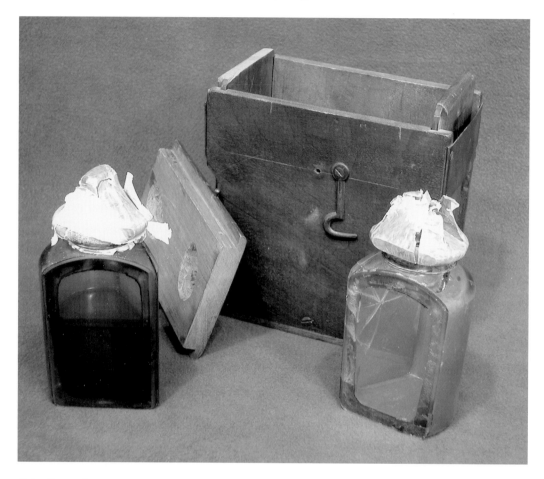

John Browning
London, England
About 1870
Height: 5 inches
Exhibited in "Illuminating Instruments," Hood Museum of Art, Dartmouth College, 2004.   2002.1.34499

By varying the composition of glass, generally by the addition of lead salts, different refractive indices can be obtained; the higher the refractive index, the greater will be the dispersion of white light. However, for some applications a refractive index is needed that cannot be achieved through the use of glass. These are examples of bottles, originally filled with carbon bisulfide, that act as high-refractive-index prisms. Carbon bisulfide has an index of refraction greater than glass, and the sides of the prisms do not have any effect because their inner and outer surfaces are parallel.

These prisms are hand signed, *John Browning, London.* They are capped with a multiple-layer seal to prevent evaporation and, indeed, one is still half full. A late-nineteenth-century inventory describes these prisms as "for spectrum projections." Professor Charles A. Young visited John Browning's shop in January 1871 and probably purchased these prisms then.[55]

SPECTROSCOPE

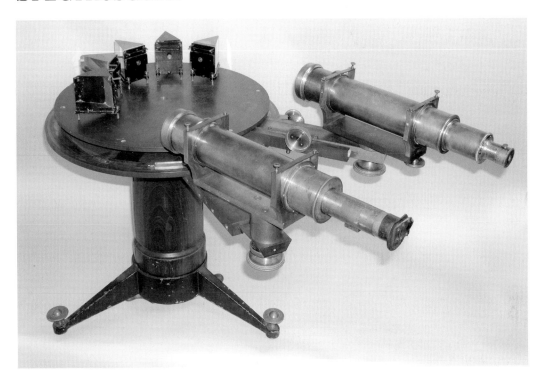

Unsigned (Alvan Clark & Sons)
Cambridgeport, Massachusetts
1867
Prisms: 45 degrees, base 2 inches; Aperture of telescope and collimator: 2 1/4 inches;
Diameter of base: 18 inches   2002.1.35336

A spectroscope separates sunlight or other polychromatic light into its constituent wavelengths, forming a spectrum. The device consists of a system of slits and lenses (collimator) that focuses a narrow beam of parallel rays from the light source onto the refracting element (glass prisms, hollow prisms filled with carbon bisulfide, or diffraction grating). The refracted rays pass through another set of lenses, essentially a small telescope, through which the spectrum may be studied visually or photographed.

Early in the nineteenth century, the English chemist William Wollaston and the German optician Joseph Fraunhofer discovered hundreds of dark lines in the solar spectrum and some bright lines in chemical spectra such as candlelight. By midcentury, the Heidelberg professors, Robert Bunsen and Gustav Kirchhoff, showed that such lines yield information about the elemental composition of the light source (bright emission lines) or of any medium through which the light passes on its way to the spectroscope (dark absorption lines). During the second half of the nineteenth century, spectroscopy became a major activity in both physics and chemistry, and provided the foundation for the new discipline of astrophysics. Linking the laboratory and the telescope, spectroscopy provided the first means to investigate the chemical composition of celestial objects.

Early spectroscopists found that by increasing the resolution of their spectroscopes they could discover additional spectral features, not only more lines but also differences in lines' width, intensity, sharpness and shape. To "zoom" in on such details, spectroscopists sought to disperse the spectrum over wider angles. For his famous 1861 map of the solar spectrum, extending more than 2.5 meters in length, Kirchhoff employed a spectroscope made by C.A. Steinheil of Munich in which the light passed through a train of four flint-glass prisms. In 1863, the Harvard chemist Josiah P. Cooke, Jr. reported finding "at least ten times as many lines as are given by Kirchhoff in his chart" by using a spectroscope of nine carbon bisulfide prisms (see entry) made by Clark & Sons. The world's most powerful spectroscope to date, this multi-prism instrument employed a movable iron cone to vary the circumference of the circle against which the prisms rested, so that the observer could sweep through the spectrum with the prisms automatically adjusting

Fig. 23.—General View of Spectrum Photographic Arrangements showing Heliostat Lamp and Lenses.

to the angle of minimum deviation for any given color. Spectroscopists prefer to measure lines at that angle, because near it wavelengths are roughly proportional to the linear position of the lines.

When in 1866 Charles A. Young returned to Dartmouth as Appleton Professor of Natural Philosophy and Professor of Astronomy, he immediately took up solar spectroscopy, initiating a research trajectory that would make him by century's end one of America's most significant astronomers. After corresponding with Cooke (whose father was a Dartmouth graduate) and traveling to Boston to consult with the Clarks, Young ordered a Clark & Sons spectroscope consisting of nine dense flint-glass prisms, a calibrated steel base, and a collimator and telescope of 2 1/4 inch aperture, for which he paid $450. Lacking the Cooke-Clark mechanism for automatic adjustment, Young's first spectroscope apparently was designed for use on a laboratory bench in conjunction with a heliostat, an arrangement Kirchhoff and the British astronomer, J. Norman Lockyer, had employed (see illustration above). Only six of the original prisms, some quite chipped around the edges, remain in the King Collection; this spectroscope remained in use at Dartmouth until 1963.[56]

By 1869, Young had mounted five of the Clark prisms in a "prism box" that he attached to the 4-inch comet-seeker telescope that he and his father had purchased in 1853 from Merz for the new Shattuck Observatory. In this configuration, the prism train gave a total deviation of 165 degrees and a dispersion of about 18 degrees between the A and H lines of the solar spectrum. In July 1869, Young published his first observations of the solar chromosphere made with this device, announcing two previously unknown emission lines. One month later, Young joined a solar eclipse party to Burlington, Iowa, organized by the superintendent of the U.S. Navy's *Nautical Almanac*, and used his prism box and comet-seeker to co-discover the first emission lines seen in the solar corona (visible only during a total solar eclipse). Young interpreted one bright green coronal line, 1474K (Kirchoff's units, replaced in the 1880s by Angstrøm's

80

Accessories for the 1867 spectroscope, assembled by Allen King in 1963

units), as signifying the presence of iron vapor in the corona, a puzzling result that would prompt much speculation until its eventual explanation in 1941 as the result of rare transitions in thirteen-times ionized iron atoms (requiring a coronal temperature in excess of 1 million degrees).

Given the significance of his 1869 findings, Young was invited to participate in the U.S. Coast Survey's solar eclipse expedition to Spain and Sicily in December 1870, organized by Benjamin Peirce and funded by a grant of $30,000 from the U.S. Congress. Among others, the party included professors Joseph Winlock of Harvard and Edward C. Pickering of MIT as well as Alvan Graham Clark, one of the Clark sons (see illustration, page 86). By September of that year, Young and the Clarks had devised a new compact, light, 15-pound spectroscope. Made to attach by means of a brass coupling to the College's 6.4-inch Merz refracting telescope (see Shattuck Observatory entry, page 60), the telespectroscope was designed to analyze sunspots and solar prominences (enormous jets of incandescent gas thrown up from the surface of the Sun). Indeed, with the help of Hanover photographer, H.O. Bly, Young in the fall of 1870 made the first photograph ever of a prominence. He took this telescope and spectroscope to the eclipse site in Jerez, Spain, wanting to continue his coronal investigations. He mounted the refractor on a 25 x 25 cm wooden post, placed under a large tent borrowed from the Jerez Cricket Club (see illustration, page 86). Training the collimator slit tangent to the solar limb, Young at the first instant of totality glimpsed the "flash spectrum," a brief moment when the hundreds of Fraunhofer (dark absorption) lines reverse into bright emission lines. This observation, predicted earlier but never before seen, would be the most important discovery of the 1870 eclipse, for it seemed to demonstrate that the Fraunhofer lines are produced by absorption in the atmosphere just above the solar surface.

Young's description of the dramatic moment would be widely reprinted: "As the crescent grew narrower... the dark lines of the spectrum and the spectrum itself gradually faded away; until all at once, as suddenly

81

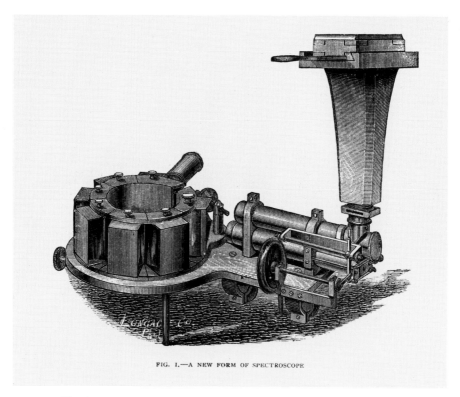

FIG. I.—A NEW FORM OF SPECTROSCOPE

Wood engraving of Young's spectroscope. Prototype used in Spain, 1870

as a bursting rocket shoots out its stars, the whole field of view was filled with bright lines more numerous than one could count. The phenomenon was so sudden, so unexpected, and so wonderfully beautiful as to force an involuntary exclamation. Gently and yet very rapidly they faded away, until to within about 2 seconds ... they had vanished." Reporting these results to the American Association for the Advancement of Science, *American Journal of Science*, *Chemical News* and the newly formed journal, *Nature* (the latter two published in London), Young's early solar spectroscopy attracted considerable attention and represented the first internationally significant scientific research conducted at Dartmouth.[57]

The original form of Young's telespectroscope (used in Spain) contained a train of six prisms plus a seventh half-prism, on the back of which was cemented a right-angle prism to reflect the light back through the upper part of the train. Such a two-pass arrangement, a novel idea first implemented by the Clarks, provided a total dispersion equivalent to 13 prisms around a compact ring only 8 inches in diameter. The collimator and telescope remain fixed. Like the earlier Cooke-Clark design, the circumference of the circle around which the prisms stand could be expanded or contracted to automatically adjust the prisms to given angles of minimum deviation. With the total amount of dispersion adjustable by placing the last prism with the reflector in any position, the Young-Clark telespectroscope was among the most powerful and convenient instruments then developed for solar spectroscopy.

In his first description of the telespectroscope published in October 1870, Young announced several modifications to make the instrument "still simpler, firmer and perfectly automatic in its adjustment" that he would introduce after the Spanish eclipse expedition. By replacing the first prism of the train with a half-prism affixed to the base with its front face perpendicular to the collimator, the remaining prisms could be hinged together and controlled by slotted radial bars sliding on a common center (visible on the illustration above), a design vaguely proposed in 1863 by the Viennese maker, Otto von Littrow, and further enhanced in 1865 by the pioneer American spectroscopist, L.M. Rutherfurd, and in 1870 by John Browning, who linked the telescope to the hinged prisms. The revised Young-Clark design (the current form of the instrument) reduces the effective dispersion from 13 to 12 prisms, but increases the precision of the automatic adjustment of the prisms. A micrometer attached to the eyepiece allows linear distances between lines to be measured.

82

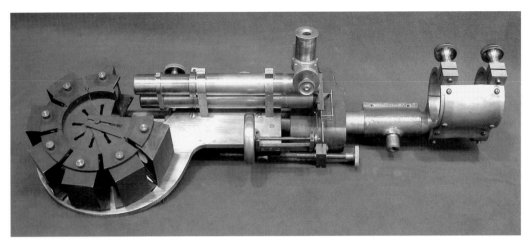

Telespectroscope
Unsigned (Alvan Clark & Sons)
Cambridgeport, Massachusetts
1870
Prisms: 55 degrees, base 1 3/8 inches; Aperture of telescope and collimater: 7/8 inch;
Diameter of base: 7 1/2 inches
Exhibited in "Illuminating Instruments," Hood Museum of Art, Dartmouth College, 2004.   2002.1.35377

Young and the Clarks also devised an oscillating mechanism, driven by the large brass flywheel visible near the center of the instrument, that was designed to rapidly open and close the slit, thereby reducing the intensity of the incoming light and enabling the solar prominences, by means of persistence of vision, to be viewed in daylight. In 1869, Lockyer and the continental astronomers, Pierre Janssen and Friedrich Zöllner, each had tried such an arrangement, but experience soon taught all these observers that the prominences could be seen more easily simply by widening the slit and leaving it stationary. Prisms of the spectroscope attenuate the continuous solar spectrum so that the outlines of the prominences, with their few bright emission lines, could be clearly seen. Young also found that the rotating flywheel "always causes a slight oscillation of the equatorial, which interferes with the definition of details, and I prefer to work with the slit simply opened."[58]

Young would use the modified telespectroscope for solar studies until 1877, when he became professor of astronomy at Princeton University, lured from Dartmouth by the promise of a new 23-inch refractor and a solar spectroscope with a diffraction grating. Indeed, the 1870 Young-Clark two-pass design signaled the zenith of the development of ever more dispersive, multi-prism telespectroscopes. Before the invention of twentieth-century optical coatings, each prism in a train reduced the intensity of the transmitted light by 10 to 25 percent. In describing his early observations of solar prominences, Young rhapsodized about "the most delicate details [which] reveal themselves with a beauty and clearness of definition which even yet always surprises me, and speaks most emphatically for the exquisite workmanship of the 43 different surfaces by which the light is either refracted or reflected on its way from the slit of the collimator to the eye." In 1871, Young even proposed making a four-pass arrangement of the prisms, suggesting that "the light of the sun is so brilliant that, in studying its spectrum, we may use as many [prisms] as we please."

By the early 1870s, John Browning of London was making two-pass, multi-prism spectroscopes, not unlike the Clarks', and advertising a four-pass model that would yield a total dispersion equivalent to 24 flint prisms. Adam Hilger of London and Howard Grubb of Dublin reputedly built six-pass solar instruments, using reflecting prisms to bend the light beam six times through a train of three prisms. Other makers experimented with compound prisms or with combining two rings of prisms in opposite directions, all seeking greater dispersion. Alvan Clark & Sons would provide multi-prism, two-pass spectroscopes to

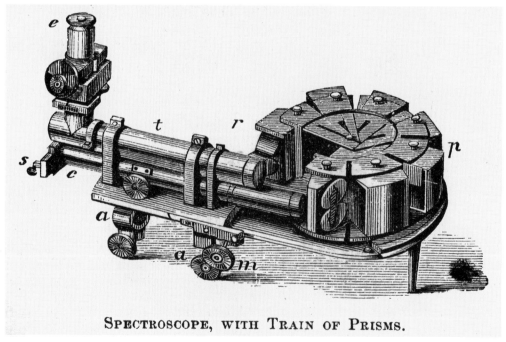

SPECTROSCOPE, WITH TRAIN OF PRISMS.

Wood engraving showing Young's redesigned spectroscope, 1881

observatories at Columbia University (1872) and Yale College (1878); a multi-pass instrument with dispersion equal to 20 flint prisms, ordered from the Clarks by Henry Draper in 1879, apparently never was completed. According to extant records, the Clarks after that date sold no further multi-prism spectroscopes.

Indeed, mechanical instabilities, the high cost of multiple prisms, and the loss of intensity in passing light through so many optical surfaces increasingly limited development of such complex spectroscopes for astronomical uses. A diffraction grating attenuates incoming light only once.

Brass coupling to connect the spectroscope to the 9.5-inch Clark refractor
Unsigned
About 1870
Length: 13.7 cm   2002.1.35377

85

Its dispersive power can be increased dramatically by reducing the spacing between the lines or turning the grating to use different orders of the spectra (see Rowland grating entry, page 95). Already in 1873, after placing a grating made by Lewis Rutherfurd into his spectroscope, Young would proclaim: "The grating is much lighter and easier to manage than a train of prisms, and if similar ruled plates can be furnished by the opticians at reasonable prices and of satisfactory quality, it would seem that for observations upon the chromosphere and prominences they might well supersede prisms." Indeed by the 1880s gratings, especially those produced in Henry A. Rowland's laboratory, would offer "a leap from the old into a new world" of spectroscopy, in the words of one contemporary observer.

As astronomical observatories seeking ever larger telescopes shifted from refractors to reflectors, the 40-inch objective made by Clark & Sons in 1897 for the Yerkes Observatory remains unsurpassed even to the present as the world's largest lens in a working telescope. Likewise, as astrophysicists seeking more dispersive power shifted from prisms to gratings, Young's telespectroscope also made by Clark & Sons may well be among most powerful multi-prism instruments ever used for significant work in solar spectroscopy. In both cases, one type of instrument design was supplanted by another. The Clarks' 40-inch lens and their multi-prism spectroscope stand as monuments to changing instrumental technologies.[59]

Eclipse expedition, Jerez, Spain, 1870. Enlarged from one-half of an albumen-print stereo view.

SMALL DIRECT-VISION PRISM SPECTROSCOPES

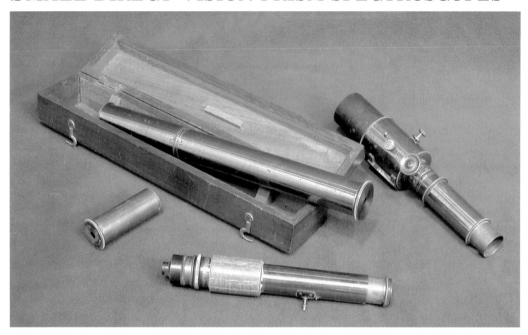

John Browning
London, England

McClean's star spectroscope (left)
About 1871
Length: 2 1/2 inches   2002.1.34488

Star spectroscope (center)
About 1871
Length: 8 inches   2002.1.34487

Grace's spectroscope (bottom)
About 1871
Length: 7.5 inches   2002.1.34486

Sorby-Browning microspectroscope (right)
About 1871
Length: 8 inches   2002.1.34485

Late in 1870 following his eclipse expedition to Jerez, Spain, Charles A. Young passed through London to meet scientists and buy instruments. These are a selection of spectroscopes he bought from John Browning. On Friday, January 27, 1871, he wrote in his diary, "at Browning's nearly all day." He visited Alfred Apps' shop on Saturday (see induction coil entry, page 160), and returned to Browning's on Monday.[60]

Small hand-held spectroscopes were popular and used to demonstrate both dispersion and spectra. Several models of small multi-prism spectroscopes were made for use with telescopes, designed to be used in place of the eyepiece. With such instruments the spectra of brighter stars can be observed visually.

Browning had a fine reputation as a maker of small spectroscopes that employ a train of in-line prisms, known as direct-vision spectroscopes (lower left). Such compound prisms, made of alternating flint- and crown-glass, produce no net deviation of a mid-spectral ray, but do disperse the spectrum. If a bright object is viewed rectilinearly through such a prism train, its spectrum will be seen in

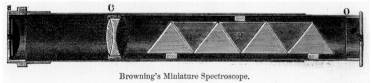

Browning's Miniature Spectroscope.

Browning has manufactured another direct-vision spectroscope, with seven prisms, which commends itself by the excellence of its performance, the facility of its use, the smallness of its di-

Fig. 49.

Browning's Miniature Spectroscope.

mensions, the purity of color, and its low price. A sketch of it is shown in Fig. 49; the slit is simply regulated by turning round a ring at the end of the tube, and the spectrum is observed direct without a telescope. The length of this admirable little instrument is only about $3\frac{1}{4}$ inches, and is, therefore, very deservedly called the miniature or pocket spectroscope.

the line of sight. The colors, however, are not as widely dispersed as would be the case were the object viewed at an oblique angle through a flint-glass prism.

The Sorby-Browning microspectroscope is a complex instrument for use with a microscope. It has an adjustable slit, a prism to permit introduction of a comparison spectrum (the instrument could also hold vials filled with various materials to study absorption spectra), and a small micrometer device near the eye end (not present on this instrument, and not shown in the wood engraving reproduced below) that could be used to measure the lines. This device, available for an additional £2 5s, provided a movable, illuminated pointer, focused on the spectrum, whose position could be read off a small micrometer and used to measure the relative positions of lines. Employing all these features makes this device somewhat fussy to use. A more robust and complex version was manufactured by Adam Hilger.[61] The Sorby-Browning microspectroscope was used to study absorption patterns of biological specimens and could be used to determine composition of liquids.

In 1878, Browning wrote and self-published "How to Work With The Spectroscope," describing his instruments and thereby advertising their availability.[62]

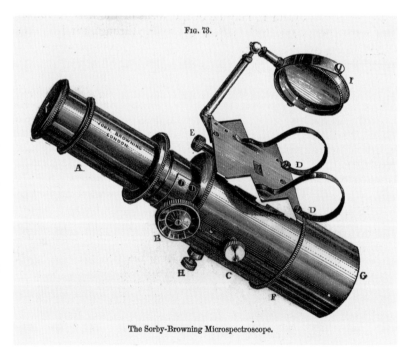

Fig. 73.

The Sorby-Browning Microspectroscope.

88

Spectra demonstrator

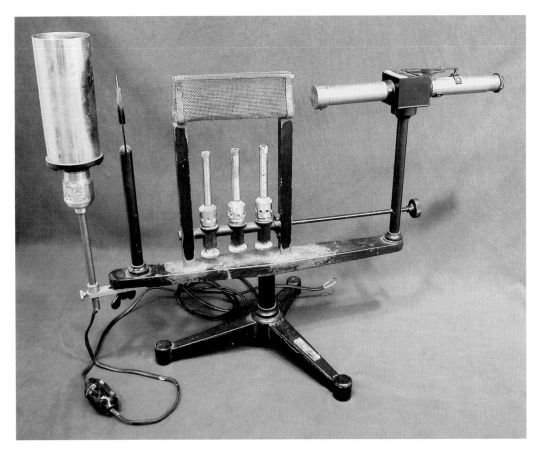

EASTERN SCIENCE SUPPLY COMPANY
Boston, Massachusetts
About 1932
Height: 36 cm; Length: 62 cm

2002.1.34773

Emission and absorption spectra, their transition from one form to another, and continuous spectra can be demonstrated using this apparatus. At the right is the observing telescope with its slit and multi-prism direct-vision spectroscope. The internal scale is illuminated by a battery-powered lamp. The gas burners in the center heat specimens to incandescence so that their spectra can be observed. A movable rod can be rotated to tilt the burners and thereby to vary the heat applied to the specimens. The incandescent, gas-filled, high-intensity electric lamp at the left provides illumination.

This instrument was in use until recently. The corrosion visible underneath the burners attests to its longevity as a demonstration instrument.

The 1932 ESSCO catalogue refers to this apparatus as "student-proof." The list price was $85; in 1935 ESSCO offered a Depression-era reduction in price to $80.[63]

90

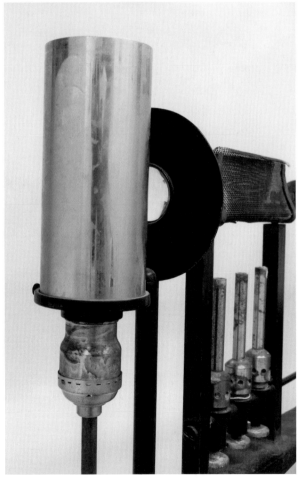

BULLETIN RR1   SEPTEMBER, 1932

ESSCO RAPID REFERENCE
PICTORIAL BULLETIN

A PICTURE—A SUMMARY—A PRICE

ESSCO-DESIGNED
UNIQUE SCIENCE EQUIPMENT

for YOUR
Lecture Room and Laboratory

ASTRONOMY—CHEMISTRY—PHYSICS
GENERAL SCIENCE

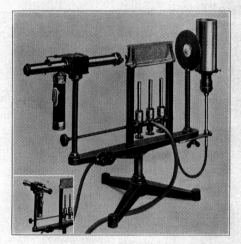

SPECTRA DEMONSTRATOR L290

Prices Subject to Change Without Notice

DESIGNED—BUILT—SOLD

by

EASTERN SCIENCE SUPPLY COMPANY

P. O. BOX 1414, BOSTON, MASS., U. S. A.

The front cover of the 1932 Eastern Science catalogue illustrates their Spectra Demonstrator. A modest publication, the September 1932 issue had 16 pages. Dartmouth faculty bought heavily from this catalogue.

HELIUM TUBE

THOMAS TYRER & CO. LTD.
London, England
1906
7.5 x 7 x 24.5 cm

2002.1.34991

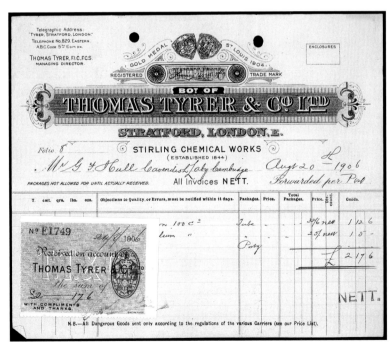

Sir Wm. Ramsay's Instructions.

This tube contains about 100 cub. cents. of **HELIUM**, at atmospheric pressure and temperature.

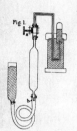

TO GET THE GAS OUT OF THE TUBE, make two cuts with a file at *a* and *b*, over each end slip a piece of thick walled india rubber tubing of about this diameter **O** The top piece should be about 2 inches or 5 cms. in length, and should be wired with copper wire at one of the constrictions of the glass tube. The glass tube should be clamped in a vertical position, and the india rubber tube filled with mercury. A piece of glass tubing of thermometer bore should be bent into the form shown in the figure, and pressed into the open end of the india rubber tube. The mercury will flow through the capillary bore and fill the capillary tube. Reverse the position of the tube in the clamp, and attach a piece of similar thick-walled india rubber tubing, about two feet or 600 cms. long to the other end, wiring it as before; wire to its other end an open reservoir, as in the sketch. Pour some mercury into the reservoir and pinch the rubber tube so as to expel all air, repeating until no more bubbles escape. Then reverse the apparatus in the clamp, so as to stand as shown in the cut. Place under the inverted siphon a mercury trough, as shown; fill a small tube with mercury and invert it over the open end of the siphon. Place a screw pinch-cock at *a*. Now break the filed ends of the helium tube through the rubber tubing, by help of a pair of pliers, the bottom one first. The apparatus is now a small gasometer, and by raising the reservoir helium can be forced out into tubes placed to receive it, the rate of passage being regulated by the pinch-cock.

TUBES SOLD BY
THOMAS TYRER & CO., Ltd., Chemical Manufacturers, STRATFORD, LONDON, ENGLAND.

This vial contained about 100 cc of helium. The accompanying sheet provides instructions for transferring the gas to another vessel. In 1905-6, Dartmouth physics professor Gordon Ferrie Hull spent a year at the Cavendish Laboratory at Cambridge where he conducted research on spectral lines. The invoice indicates that Hull bought tubes of helium and argon. In the published results of his Cavendish research, "Investigation of the influence of electrical fields upon spectral lines," Hull noted that the helium "obtained from Tyrer & Co., London, proved to be quite pure."[64]

Helium was the first element to be discovered through the use of the spectroscope. In 1868, J. Norman Lockyer identified the element in the solar spectrum, hence the name. William Ramsey in 1895 first isolated terrestrial helium and showed that the element is an inert gas. For his work on the noble gases, Ramsey won the 1904 Nobel Prize for chemistry and was knighted.

ROWLAND DIFFRACTION GRATING

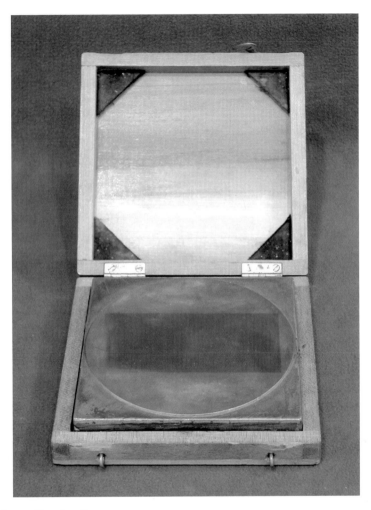

Unsigned (Henry Rowland)
Baltimore, Maryland
1885
Speculum metal flat: 0.625 x 5 x 5 inches; ruling: 4 x 1.5 inches
Exhibited in "Illuminating Instruments," Hood Museum of Art, Dartmouth College, 2004.   2002.1.34521

   The more the spectrum is "spread out" without making it too faint to study, the more detail can be seen. Instead of using prisms to disperse the light, closely ruled parallel lines called gratings also produce spectra. First developed by Joseph Fraunhofer in the 1820s, gratings are more efficient than prisms and allow absolute measurement of wavelengths. Although expensive and difficult to make, by the end of the nineteenth century gratings replaced prisms in many applications. For several decades gratings made by Henry Rowland, the first professor of physics at the new Johns Hopkins University, were the best in the world.[65]

   Louis Bell (Dartmouth, 1884) used this grating in his thesis research under Rowland, on the absolute wavelength of light. This is Bell's Number III grating, ruled in April 1885 "at a very nearly constant temperature of 10^0 C. It contains 29,000 spaces.... It is a phenomenal grating both in its superb definition and extraordinary regularity of ruling, and was selected by a large number because of its very unusual

This is an original Rowland grating and was used by Dr. Louis Bell '85 in measuring the wave length of sodium line 5896. For a very brief time it is the value of the wave length was a world standard.

The grating was used by Prof. Wm Rogers of Colby as a standard. It was acquired by me in 1898 and was brought here by me 1899–1900.

G.F. Hull

perfection. The focus of the spectra on either side of the normal is the same and the ruling is flawless."[66] For a very brief period of time, Bell's value of the sodium wavelength was the world standard.

Gordon Ferrie Hull, who in 1897 had completed his Ph.D. under Albert A. Michelson, acquired this grating in 1898 from Professor William Rogers of Colby College, so that Hull could make his own measurements of the wavelength of light.[67] From 1899 to 1940 Hull taught physics at Dartmouth College.

This grating is ruled on a flat, polished square of speculum metal, a highly reflective alloy consisting mainly of copper and tin. The metal flat was provided by John Brashear of Pittsburgh and although it was usual for both Brashear and Rowland to sign the gratings, this one is unsigned. The wooden box is marked *Absolute Wave Grating III*.

Although this grating is flat, Rowland also succeeded in ruling concave gratings, where the lines are perpendicular to the long axis of a concave, cylindrical, polished mirror. Such a grating requires no focusing lenses and obviates the reflection and absorption they produce.

96

RULING ENGINE

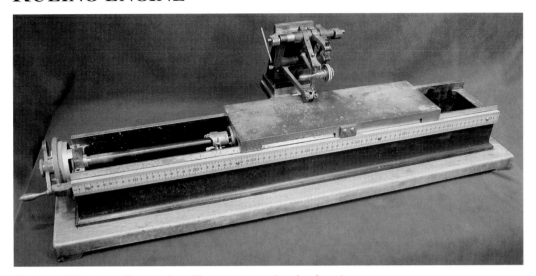

Unsigned (Société Genevoise d'instruments de physique)
Geneva, Switzerland
1899
15 x 91 x 21.5 cm
Exhibited in "Illuminating Instruments," Hood Museum of Art, Dartmouth College, 2004.   2002.1.35237

This hand-operated screw and bed was used initially for graduating scales and then calibrating them against standard lengths. Fifty years later it was modified for ruling diffraction gratings. The King Collection has examples of several unsuccessful attempts to produce a grating, a small disk bearing a faintly lined grating dated January 12, 1936, and a large flat on which are ruled several gratings labeled "First ruled at Dartmouth," but dated August 31, 1936. This instrument was also used in E.F. Nichols' and G.F. Hull's 1903 experiment to measure the "pressure of light."

The lead screw, 40 cm long, has a calibrated wheel permitting linear travel of 0.01 mm to be read directly. The tracelet remains stationary so that transversals as well as lines perpendicular to the axis can be made.

The Société Genevoise catalogue of 1900 presentsan engraving of a ruling engine that appears nearly identical to our machine, priced at SF650.[68]

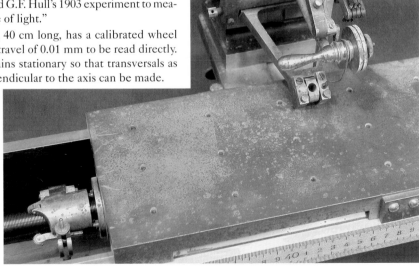

REFRACTING STEREOSCOPE

Duboscq-Soleil
Paris, France
About 1853
15 x 18 x 11 cm

2002.1.34544

   The stereoscope, invented by Charles Wheatstone in the 1830s, is an optical device that superimposes images from two images of the same subject recorded at slightly different angles, creating the illusion of depth, or "3-D." The device shown here, the lenticular or refracting form of stereo viewer, was invented by David Brewster, who announced his invention in 1849. After being unable to find a British manufacturer, Brewster went to Paris, where the first stereo viewers of his design were made by Duboscq-Soleil and exhibited at the 1851 Great Exhibition. The device created a sensation, interesting even Queen Victoria.

   Stereo viewing rapidly became a parlor pastime that approached the level of a craze, especially after the introduction in the late 1850s of the familiar stereo card, containing two images. The development of the stereoscopic camera permitted stereo images to be produced, first by the calotype process, then by ambrotype, followed by albumen prints. Toward the end of the craze, stereo cards were being printed rather than using the actual photographic prints. The Dartmouth instrument illustrated here (purchased for $4) is a fine, early example of a consumer item manufactured by a noted instrument firm.[69]

   Our stereoscope is covered with mahogany veneer, lined with black felt, and missing the front panel. It bears French patent markings: *Breveté s[ans] G[arantie] du G[ouvernemen]t.*

98

Opthalmotrope

*CAMBRIDGE SCIENTIFIC INSTRU-
MENT CO LTD*
Cambridge, England
About 1920
Length: 15 cm
Serial number: 3427   2002.1.34574

This mechanical representation of the eye musculature is one of the few models in the King Collection. The "eyes" are painted wood; a brass ball joint on their posterior aspects permits a wide range of motion. The oculomotor muscles are represented by springs; their complex action is mimicked by the addition of an elaborate network of linkages.

Many of the leading nineteenth-century sensory physiologists developed opthalmotropes. In 1846, Th. Ruete coined the term and constructed the first model to demonstrate eye rotation. Over the next fifty years, F.C. Donders, H. Knapp, E. Landolt, A. Graefe, H.P. Bowditch, H. Aubert, and W. Browning all published schemes for opthalmotropes. Perhaps the most sophisticated model was proposed in 1862 by W. Wundt, the Heidelberg physiologist. By adjusting both weights and springs to model the cross-sectional area and length of the muscles, Wundt sought to represent the physical forces in ocular motion.

The 1899 catalogue of the Cambridge Scientific Company offers an eye model for £2 10s, nearly identical to ours, that is attributed to Thomas Peter Anderson Stuart, professor of physiology at the University of Sydney since 1882. However, none of the standard opthalmological works from around 1900 mention Stuart in connection with opthalmoscopes, and none of his known publications mention the model. It is quite possible that this trade catalogue attribution is not accurate. For example, this catalogue also attributes an opthalmotrope, known to be designed by Knapp, to Helmholtz.[70]

MICROSCOPE LAMP

R & J. BECK
London, England
1884
Height: 34.3 cm
2002.1.34599

Deflecting sunlight through a specimen by means of a microscope substage mirror produces bright, uniform illumination, but is dependent upon visibility of the Sun. Hence numerous artificial light sources were designed to provide reliable illumination.

In order to achieve maximum resolution, the light entering a microscope must be able to satisfy a number of criteria. In this device, the light from the oil-burning wick can be concentrated and focused by means of a movable condenser lens. Further adjustment of the light would have been made using the substage condenser on the microscope. This lamp has many adjusting mechanisms, ranging from simple joints to precision ways, making it very versatile.

In 1878 R & J. Beck opened an optical shop in Philadelphia, which offered "microscopes, telescopes, opera glasses, spectacles, eye-glasses, photographers outfits for amateurs." A local photographer, William H. Walmsley, who earlier had been selling prepared microscope slides, soon became manager of the establishment. By 1884,

JOSEPH BECK.   ROBERT KEMP.   CHARLES COPPOCK.

AN ILLUSTRATED CATALOGUE

OF

MICROSCOPES AND OTHER SCIENTIFIC INSTRUMENTS,

MANUFACTURED BY

R. & J. BECK,

LONDON.

AND

No. 921 CHESTNUT STREET,

PENN MUTUAL INSURANCE COMPANY'S BUILDING,

PHILADELPHIA.

W. H. WALMSLEY, Manager.

— PRICE TEN CENTS. —

Among the complex adjustments is this slotted arm that serves to alter the position of the way holding the mirror.

102

the firm had become "W.H. Walmsley & Co., successors to and sole agents for R. & J. Beck," offering apparatus inscribed with both names. It issued several catalogues of optical and (increasingly) photographic apparatus before disappearing from the city directories in 1894. Walmsley also developed an "improved" photomicrographic camera and exhibited his photomicrographs at several national exhibitions.[71]

ZENTMAYER MICROSCOPES

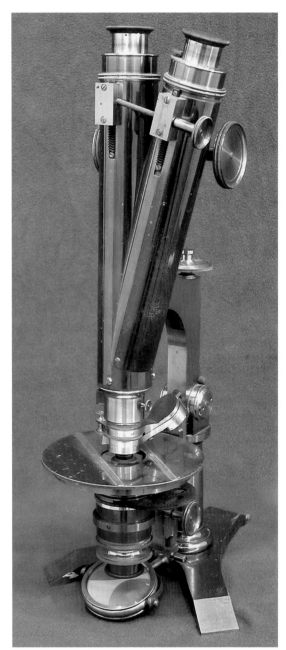

Joseph Zentmayer was a German immigrant who had settled in Philadelphia and was apparently the third commercial microscope maker (1854) in the United States, preceded only by C.A. Spencer and William and Julius Grunow. Zentmayer devised the swinging substage condenser to illuminate both opaque and transparent specimens (U.S. patent 181,120).

The King Collection has two similar Zentmayer binocular microscopes, both given to the collection by the Geology Department. These microscopes utilize Wenham prisms to split the image, providing binocular vision. The interocular distance is adjustable by the screw seen between the two tubes.

Both instruments are fitted with a rotating stage and Nicol polarizer/analyzer prisms permitting the detection and analysis of birefringence (see entry, page 112). The polarizer prism fits over the bottom end of the substage condenser and is visible in the next illustration. The analyzer is contained in a small brass cylinder that screws between the nosepiece and the objective (thereby also changing tube length). A knurled ring allows the operator to

J. Zentmayer
Philadelphia, Pennsylvania
About 1880 (Patented 1876)
Height: about 44 cm high; Diameter rotating stage: 11.2 cm
Serial number: 2424

Gift of Mrs. Moses Dyer Carbee, Haverhill, New Hampshire, 1929.

2002.1.34297

rotate the analyzer prism. A turned brass cover screws onto the end of the prism holder when not in use.

The substage condenser (below) does not contain an iris diaphragm, but rather a series of aperture stops. Several of these have central occulting discs, which would permit the microscope to give "dark field" illumination. Despite the presence of the rotating stage and rotating prisms, there is no facility for centering the optics to make the optical axis coincident with the mechanical axis.

Although both microscopes are similar, they do display several significant differences, especially in finish, all indicating that 5367 was a "deluxe" version. For example, an owner's name, *John Dexter Locke* (whom we are unable to identify), is engraved on the barrel of 5367, and the top surface of its rotating stage has been decorated. This serpentine pattern, produced by a hand scraper, has no function other than being decorative and is reminiscent of similar ornamentation found on the seismograph (see entry, page 19). The King Collection's Crouch microscope (2002.1.34296) has a similarly decorated stage and the flat top of the Steinheil spectroscope (2002.1.35419) also has such frosting.

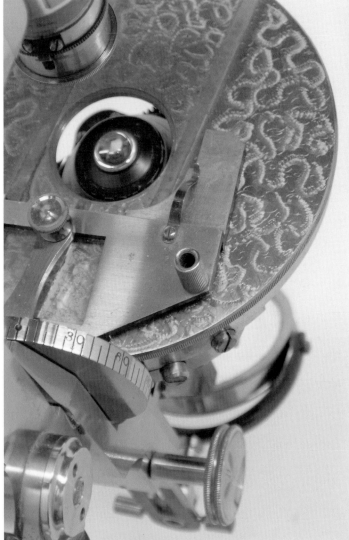

J. Zentmayer
Serial number: 5367
Gift of Mrs. Moses Dyer
Carbee, Haverhill, New
Hampshire, 1929.

2002.1.34298

Only 5367 has a degree indicator, seen on the left, and an associated pointer connected to the swinging substage condenser. Present also is a mechanical slide holder to facilitate movement of specimens. The rotating nosepiece holds four objectives, whereas in 2424 the nosepiece holds only two.

The swinging condenser rotates to extreme positions, including above the stage as seen in the final image below. Here, with the objectives removed for the sake of clarity, we can see the condenser situated to focus light onto, rather than through, a specimen. The capability of such extreme movement makes it imperative that the light source be equally flexible and able to be pointed in many directions. The complexity of the light source shown in the previous example may well be due to such requirements.

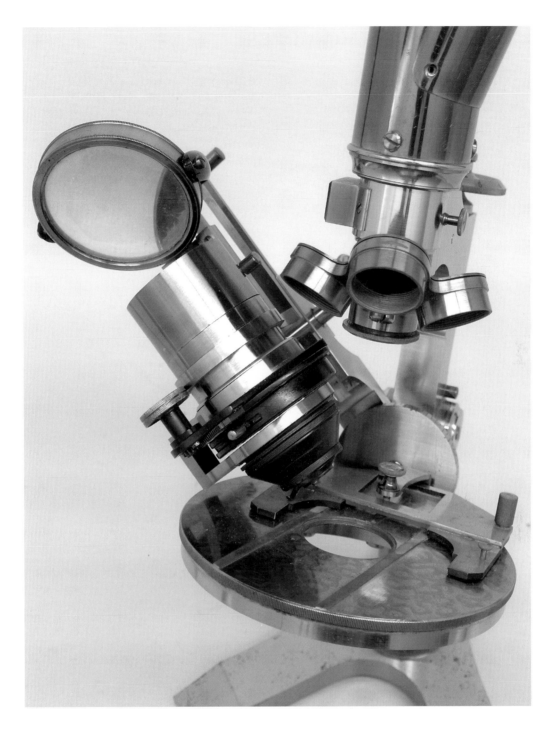

Andrew Ross of London incorporated the Zentmayer swinging substage condenser into his compound binocular microscope in 1876, thereby providing an interesting case of an American instrument design being transported to London. The instrument became known as the Ross-Zentmayer microscope.[72]

PROJECTING MICROSCOPES

Unsigned (Jules Duboscq)
Paris, France
1853
Approximately 26 x 23 x 23 cm

2002.1.34305

Light emitted by an oil-burning lamp (see entry, page 100) generally would have been sufficient to illuminate the specimen, either by reflected or transmitted light, on a conventional microscope viewed through the ocular by a single individual. In order to demonstrate the microscope image simultaneously to several people, the image must be projected or split. A projected image requires a more intense light than can be generated by conventional oil-burning lamps.

Like the magic lanterns of the seventeenth century, the solar microscope uses the sun as a source of light. A mirror on the outside of the building reflects sunlight into the microscope with sufficient intensity to project an image in a darkened room. The Dartmouth instrument is incomplete, and consists largely of the coelostat and attachments, shown above. The Duboscq catalogues refer to this device as a *port lumière*. The 1862 inventory of Dartmouth's Appleton apparatus lists the value of the above device, with condensing lens, at $55; the shutter (now missing) was valued at $2.[73]

Projection microscopy played an important role in teaching, since the same specimen could be viewed by a large number of students while it was being demonstrated by the instructor. However, the size of the projected image and the magnification possible was still restricted by the intensity of the light source. In the twentieth century, a variety of light sources were used, including mercury arc lamps, but it was discovered that, as the light became brighter, it had increasingly deleterious effects on the specimen being viewed. Stained sections often suffered photobleaching, causing many microscopists to call such devices

108

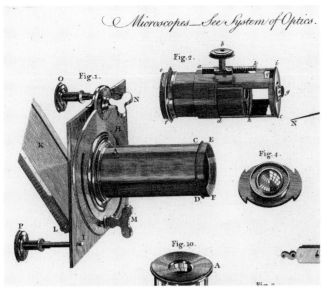
Engraving of solar microscope, 1789

"microincinerators," in an attempt at amusement.

The Leitz microscope below represents the last generation of optical projection microscopes and employs a large xenon light source. The operator's eyes are protected by a flap of dense filter and goggles. Magnification is changed by sliding different objectives into place. The image is projected onto a screen by the periscope at the top. This instrument was in use at Dartmouth until replaced by digital imaging techniques around 2000.

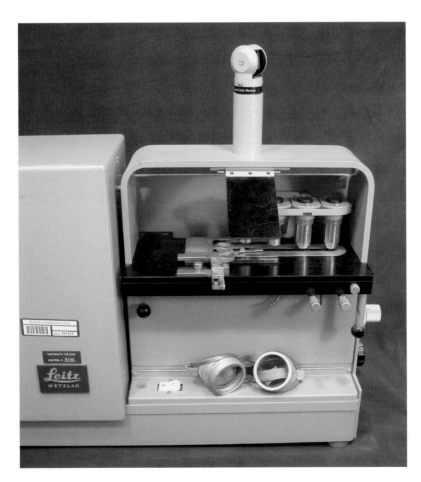

Leitz
Wetzlar, Germany
About 1970
66 x 77 x 26 cm
(Shown without separate power source for light)
2002.1.35230

POLARIMETER WITH BIREFRINGENT SPECIMENS

J. Duboscq-Soleil
Paris, France
1853
Height: 57 cm (22.4 inches)
Exhibited in "Illuminating Instruments," Hood Museum of Art, Dartmouth College, 2004.   2002.1.34459

   A polarimeter detects and measures the rotation of a polarized beam of light passing through an optically active material. Light generally vibrates in all planes. If we somehow remove all the light waves except those vibrating in one plane, we have polarized light. Some naturally occurring crystals will polarize light passing through them; light also becomes polarized by reflection.

   J.B.F. Soleil's polarimeter, invented about 1845, polarizes light by reflection from multiple layers of glass (eleven layers in this instrument, movable from 0^0 to 90^0). A second stack of glass plates acts as an analyzer (an analyzer is simply another polarizer that, when rotated, will alternately block or pass the polarized light through the polarizer). Substances to be studied are placed between the polarizer and analyzer. The discs contain cork-mounted crystals. The sample holder, maintaining the specimen in the light path, can be moved along several axes.

   J.B.F. Soleil and his son-in-law, Jules Duboscq, were the premier optical makers in Paris in the nineteenth century. Charles and Ira Young made several visits to their shop in 1853, where they spent $330, the largest of any of their purchases in Paris. We get a glimpse of Duboscq's renowned demonstrations through Charles' diary entry: "This evening at Mr. Duboscq's I have been seeing the finest experiments I ever saw.... His experiments were so many of them magnificent, especially in polarization."[74]

110

GLASS PRESS

Duboscq Soleil
Paris, France
1853
23.5 cm

2002.1.34470

Ordinary glass and many other transparent solids are optically isotropic, meaning that their indices of refraction are equal in all directions. When physically stressed, however, glass exhibits the phenomenon of birefringence. Light passing through strained glass is broken into two rays, each traveling at different velocities and polarized with vibration directions mutually perpendicular. That is, strained glass acts like birefringent crystals such as Iceland spar. Not only optical, but also acoustical, electrical, magnetic, and thermal phenomena become direction-dependent in such anisotropic materials. Although Erasmus Bartholin had discovered birefringence in 1669, the broad study of anisotropy did not flourish until the nineteenth century.

When polarized white light passes through stressed glass and then through a second polarizing filter, colored stress patterns appear, created by the differing velocities of the two rays in the glass. The spacing between the colored bands is correlated to the amount of stress in the glass; the greater the stress, the closer the bands.

This simple brass thumbscrew compresses a deformable substance held in its jaws (capable of holding a specimen up to about 2 x 2.5 cm). When placed in a polarimeter (see previous entry), the glass press enables the phenomenon of birefringence to be studied.

Nicol prisms

Unsigned pair of large prisms (Frans Schmidt & Haensch)
Berlin, Germany
1871
7.6 x 15.25 cm   2002.1.34498

Unsigned (Julius Grunow?)
New Haven, Connecticut
1862
(lower left) 4 x 3 cm; (right front) 3.7 x 5 cm; (right rear) 2.6 x 5 cm   2002.1.35380

Anisotropic crystals, such as those of quartz or calcite, have the property of double refraction (birefringence). A Nicol prism consists of two doubly-refracting prisms, ground and polished from such crystals, with appropriate angles, and cemented together. One of the refracted rays is able to pass through the cemented pair while the other is reflected out the side. Thus, a Nicol prism produces a single beam of light vibrating in one plane only. William Nicol of Edinburgh developed this device in 1828.

Large Nicol prisms are difficult to make because of the large size of crystal required and consequently, they are expensive. Charles A. Young paid RM80 ($65) for these two large Nicol prisms. Smaller prisms can be found in a variety of instruments such as polarizing microscopes, where they serve as polarizer and analyzer.[75]

The large Nicol prisms above have with them a note: "These Prisms were repaired and polished by Bausch and Lomb, 1930. They should not be used in projection without water between them and the lantern." Apparently excessive heat from projectors used for classroom demonstrations could damage the prisms.

112

FLUORESCENT SPECIMENS

Dr. G. Grübler & Co.
Leipzig, Germany
Early twentieth century
Case size: 2 x 22 x 32 cm

2002.1.34693

When some substances are struck by light of specific wavelength, they emit their own light, usually of a wavelength slightly longer than that of the exciting light. The phenomenon is called fluorescence and is the basis for the "black light" shows with which we are all familiar. Many naturally occurring minerals and organic materials exhibit fluorescence, but many synthesized organic molecules do so also. The latter are often used as markers in experiments and instruments that measure their fluorescence are called fluorimeters.

This is a remarkable set of eighty different fluorescent organic compounds, each in a labeled small glass vial with a stopper and sealed with red sealing wax. The interior of the case is lined with well-worn plush velvet. When excited by ultraviolet light, most of the specimens still fluoresce.

In 1880 Georg Grübler, a pharmacist trained at Carl Ludwig's famous physiological institute at Leipzig, opened a private laboratory in which he tested the suitability of dyes, obtained from the new German dye industry, for staining biological tissues. Those dyes found to be acceptable were purchased in bulk, repackaged, and sold to biologists and histologists. In 1897 the retail part of the firm became known as Dr. G. Grübler & Co., Farben, chemische Präparate für Mikroskopie und Bakteriologie. In the same year, Grübler sold the retail shop to Karl Hollborn, who in 1905 sold it to Johannes Schmid. Nonetheless, the products remained known until well into the twentieth century as "Grübler dyes." [76]

SPHEROMETERS

J. Duboscq-Soleil (disk), *Roger Matheiu* (screw), *Rigault* (glass dome and wooden base)
Paris, France
1853
Height: 11 cm
Circular scale reads 0-500; vertical reads 60-120   2002.1.34464

E.G. Smith
Early twentieth century
Columbia, Pennsylvania
Height: 9 cm
Circular scale reads 0-100; vertical reads 0-40
Both exhibited in "Illuminating Instruments," Hood Museum of Art, Dartmouth College, 2004.   2002.1.34426

   Spherometers are used to measure the curvature of optical surfaces. The instruments shown here are the most common form. They rest on three radial points and have a micrometer screw in the center. The sagitta or depth of a surface curve can be determined by zeroing the instrument on an optical flat, then measuring the curvature of an unknown optical surface by means of the central micrometer. If the sagitta and the diameter of the legs are known, the radius of curvature and hence the focal length can be calculated. Since the equation to determine the radius of curvature requires the square of the radius of the spherometer, modern ones are often made so that this number is a whole, for purposes of easy calculation. This is not the case for either spherometer shown here. The Duboscq-Soleil instrument has a measured radius of 43.6 mm. An instrument as sensitive as this one can also be used to determine the uniformity of the curvature of an optical surface by measuring at different points.

114

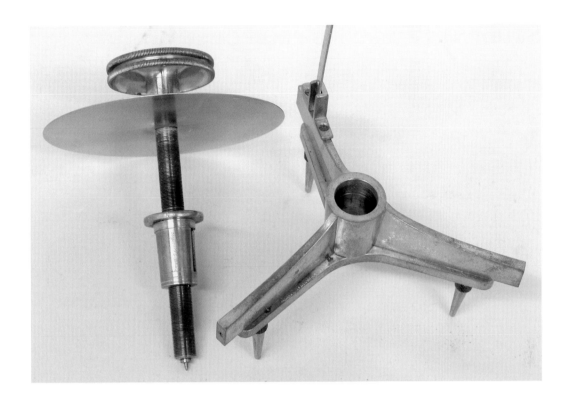

The spherometer on the left is a small student version, rather crudely made. Its tips do not end in sharp points, but rather have a diameter of about 1.1 mm. This would be sufficient to introduce significant error in use, but more serious is the fact that each of the three tips is a different distance from the center micrometer screw. The respective radii on the instrument illustrated are 26.6, 28.1, and 27.4 mm. Dartmouth has an additional three spherometers, seemingly identical, made by the same maker and exhibiting similarly disparate radii. Clearly E.G. Smith cannot be included among the precision instrument makers.

The vertical scale on the Duboscq-Soleil instrument appears to be a later replacement. An interesting construction feature of this instrument is the way the micrometer is held in place. The precision thread runs in a sleeve, seen above, that is a slide fit in the base. It is not clamped in place. Thus, if the screw is turned beyond contact with the glass, the entire assembly rises and does not harm the screw, tip, or the glass.

This instrument, purchased by Charles and Ira Young on their trip to Europe in 1853, carries some intriguing information.[77] To the bottom of the wooden base is affixed a printed paper label signed by glassblower *Rigault, Bombeur de verres, fabricant, Rue Guénégaud, No. 16, Paris*. The precision screw is roughly engraved *roger Matheiu, r. S. Severin 7, 1853*. The Duboscq-Soleil enterprise was a large one and clearly used components made by other specialized shops.

115

STUDENT TELESCOPE PROJECT

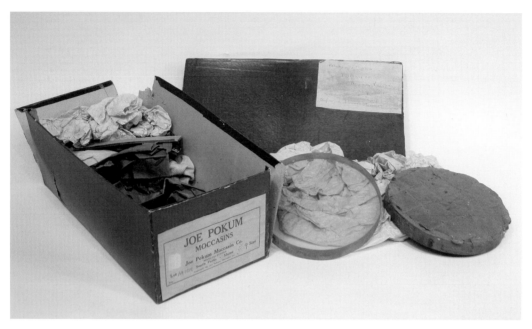

Norman D. Peschko (signed on box)
Hanover, New Hampshire
About 1937
Diameter: 6 inches
Exhibited in "Illuminating Instruments," Hood Museum of Art, Dartmouth College, 2004.   2002.1.34791

In the early decades of the twentieth century, the making of reflecting telescopes by amateurs became widespread in the United States, primarily as a result of the efforts of Albert Ingalls, an associate editor at the *Scientific American*, and a small group of enthusiasts located a few miles south of Hanover, in Springfield, Vermont, then a center of precision machine-tool manufacturing. The most commonly made instrument had a 6-inch f/8 parabolic primary mirror and was configured as a Newtonian telescope.[78]

A similar, partially finished 6-inch mirror shown above was being worked by a Dartmouth undergraduate, Norman D. Peschko, Dartmouth 1937. The plate glass mirror is accompanied by a square-faceted, rouge-stained pitch lap used for polishing and figuring. This simple project is in the tradition of telescope mirror making since the time of Newton.

Peschko's project illustrates the point that, using extremely simple methods employing readily available materials, it is possible to fashion optical surfaces of extreme precision, accurate to a fraction of the wavelength of light. No elaborate machines are necessary (unless the mirror is so heavy that hand work becomes impossible); all the great lenses and mirrors of the world's telescopes were produced by similar methods.

Peschko, who would become a research chemist, apparently did not start this mirror as part of a class project but most likely did it on his own. According to his Dartmouth transcript, he had never enrolled in an astronomy course.

Another telescope project by a Dartmouth student yielded a rather different result. In 1935-36, John W. Lovely, Dartmouth 1937, enrolled in Professor Richard Goddard's introductory astronomy class. A native of Springfield, Vermont, Lovely successfully built a 6-inch reflector for the course. With the support of his father, a major figure in Springfield's machine-tool industry, Lovely went on to construct an 8-inch reflector described more fully by Professor Goddard on the next page. The tube and pier of Lovely's telescope remain, but its mounting, mirror and spider appear to be missing.

116

EIGHT-INCH REFLECTING TELESCOPE

New Telescope

OBSERVATORY ADDS 8-INCH REFLEC-
TOR MADE BY JOHN W. LOVELY '37

THE Shattuck Observatory has a new eight-inch reflecting telescope. It came about as a result of the hobby of John W. Lovely '37. While he was taking Astronomy 2 in his junior year he completed a six-inch reflector which he brought to class and used for the remainder of the course. He then offered to make an eight-inch reflector for the Observatory. It should be noted in passing that the nation-wide hobby of telescope making had its origin in Springfield, Vermont, and that John Lovely hails from there.

By the end of the summer he had finished the grinding and polishing of the parabolic mirror for the telescope and went on to build a skillfully designed equatorial mounting. During his senior year he devoted his vacation time to work on the telescope bringing it to completion at the time of his graduation from College.

Equatorial telescopes are usually equipped with a driving mechanism which makes them follow the celestial object under observation in its diurnal motion across the sky. John considered the use of a synchronous motor for the drive of this telescope, but found that graduate work at M.I.T. kept him so completely occupied that he had no opportunity to work out the details of the scheme. At this point

John's father, Mr. John E. Lovely, vice president and chief engineer of the Jones and Lamson Machine Company, decided that the motor drive problem appealed to him as a winter's hobby for himself. Accordingly the eight-inch reflector continued its growth in the expert hands of Mr. Lovely. By early spring of 1938 the telescope was finished and ready for use.

PROPER HOUSING DESIGNED

So delicate a mechanism required a suitable housing to give commensurable utilization. After discussion of the projected uses of the new instrument, Mr. Lovely accepted the responsibility of designing the proper building for it. Before late Autumn a third white brick unit had appeared on Observatory Hill. As the pictures show, its roof may be rolled away toward the east to uncover the eight-inch reflector and its twelve foot square observing platform. The telescope is mounted on a concrete pier which is set on bed rock. The telescope is a Newtonian type of five-foot focal length with the eyepiece carried on a rotating sleeve at the upper end of the tube. The height of the observing platform is such that the observer can easily reach the eyepiece without the use of a ladder, regardless of his size or of the position of the heavenly body under observation. The effect of this arrangement is to greatly increase the time actually spent in observing during a class period. Moreover, since the rolling roof exposes

the entire sky to view the student may see at a glance the whole region toward which the telescope is directed, a condition not possible in the usual dome with its small shutter opening. In practice it has been found that while a student awaits his turn to use the telescope he increases his familiarity with the visible constellations, quite as a matter of course, since the whole sky is exposed to him.

This instrument is a particularly desirable addition to the equipment of the Observatory because it furnishes an excellent example of a type of telescope which has become increasingly important in recent years at the major research observatories. Stimulating newspaper and magazine articles on the famous 100-inch Mt. Wilson reflector and the 200-inch for Mt. Palomar, now under construction, have greatly increased general interest in reflecting telescopes. Undergraduates are quite properly interested in seeing and using a reflector along with the older standby, the refracting telescope. As a result of this gift by the Lovely family and the provision of an adequate housing by the Trustees of the College, it is no longer necessary at the Shattuck Observatory to resort to a blackboard diagram to explain the essential features of the world's largest telescopes. Designed particularly for instructional use, the instrument adds significantly to the Observatory's complement of teaching apparatus.   RICHARD H. GODDARD '20.

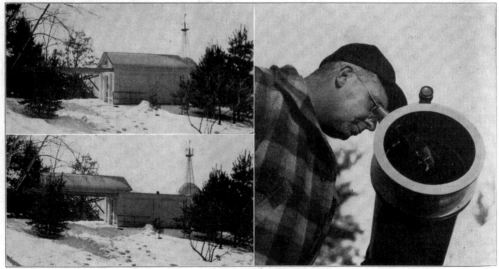

LATEST ACQUISITION OF DARTMOUTH'S SHATTUCK OBSERVATORY

The two photographs at the left show the new building housing the 8-inch reflector telescope made and presented by John W. Lovely '37, the lower picture showing the roof rolled back to expose the sky. At the right, Prof. Richard H. Goddard '20, director of the Observatory, looks through the new telescope, for which Mr. John E. Lovely, Springfield, Vt., engineer, made rotating machinery and designed the housing shown at the left.

KOLBE-HARCOURT OPTICAL DISC

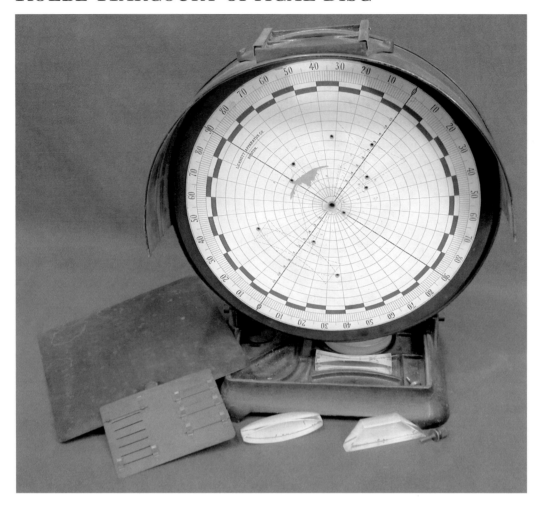

L.E. KNOTT APPARATUS CO.
Boston, Massachusetts
About 1916
Height: 42.5 cm; Diameter of dial: 30 cm   2002.1.34420

   This apparatus demonstrates the modification of a beam of light in passing through prisms, lenses, and
other optical elements. It has a heavy iron stand and a paper dial. The outer tin frame rotates and has a
counterweight for return. The base provides storage for the optical elements, that can be mounted on the
dial. Included is a hollow prism that can be filled with liquids of different refractive indices.

   Knott's naming of this instrument is somewhat peculiar. Their 1916 catalogue labels it a "Kolbe-Harcourt
optical disc." An earlier catalogue (1912) refers simply to "Kolbe's optical disc" and portrays a device of
slightly different design (a tripod rather than rectangular base). These names may reflect changing
circumstances of the company. In 1895, Louie E. Knott and Eleazar Cate founded the L.E. Knott Apparatus
Company at 16 Ashburton Place in Boston, later acquiring the Boston firm, Ziegler Electric Company, and
the scientific instrument line of E.S. Ritchie & Sons. In 1904, Knott moved to Harcourt Street in Boston's
Back Bay, where they remained until 1915, when under the sole proprietorship of Cate the company
moved to Cambridge. Knott's post-1915 catalogues designate a few miscellaneous items, such as the optical

disc, as "Harcourt" apparatus, referring perhaps to instruments having been manufactured at the Back Bay location. In 1930, Knott was acquired by Central Scientific Company of Chicago.

The eponomy in Knott's name for the apparatus is also curious. In 1896, two well-known inventors of lecture demonstration apparatus—a professor at the Viennese military geodetic institute, Hans Hartl, and a Gymnasium teacher in St. Petersburg, Bruno Kolbe—published descriptions of optical discs. Within several years, Kolbe discs were being manufactured in St. Petersburg, Berlin, Cologne and Chemnitz (by Max Kohl). Kohl also sold Hartl discs and lavishly illustrated both versions in his massive catalogues. Interestingly, Knott's "Kolbe" disc is based on Hartl's plan. Kolbe's disc, designed specifically to demonstrate the sine law of refraction, has a ground glass disc that rotates in a metal stand and differs somewhat from Hartl's arrangement. Knott's 1912 and 1916 catalogues illustrate several demonstration set-ups for the "Kolbe" disc, images that nearly exactly copy Max Kohl's diagrams for the Hartl disc. Central Scientific Company's catalogues, from 1909 (their first) through 1936, offered this disc under its original eponym— Hartl. This disc thus illustrates an American maker copying German designs and catalogue illustrations, and idiosyncratically miscopying eponymous names.[79]

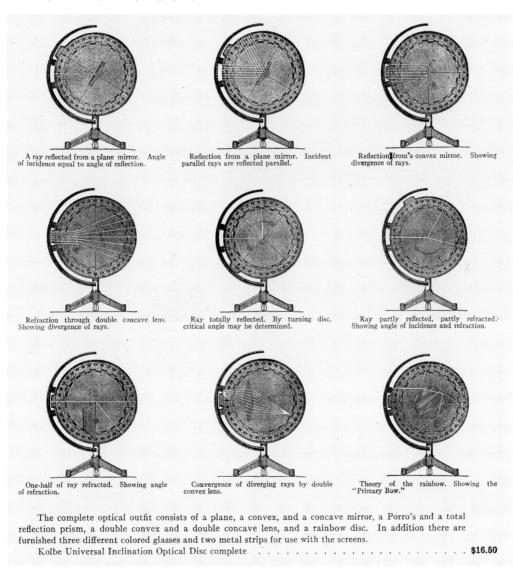

A ray reflected from a plane mirror. Angle of incidence equal to angle of reflection.

Reflection from a plane mirror. Incident parallel rays are reflected parallel.

Reflection from a convex mirror. Showing divergence of rays.

Refraction through double concave lens. Showing divergence of rays.

Ray totally reflected. By turning disc, critical angle may be determined.

Ray partly reflected, partly refracted. Showing angle of incidence and refraction.

One-half of ray refracted. Showing angle of refraction.

Convergence of diverging rays by double convex lens.

Theory of the rainbow. Showing the "Primary Bow."

The complete optical outfit consists of a plane, a convex, and a concave mirror, a Porro's and a total reflection prism, a double convex and a double concave lens, and a rainbow disc. In addition there are furnished three different colored glasses and two metal strips for use with the screens.

Kolbe Universal Inclination Optical Disc complete $16.50

Knott's 1912 catalogue

119

MICHELSON INTERFEROMETER

W. & L.E. GURLEY
Troy, New York
1908
18 x 28 cm   2002.1.34570

 The interferometer is an instrument to measure the wavelength of monochromatic light by measuring small linear distances. It splits a single beam of light into two, and compares the distance traveled by each. Upon recombining the beams, if the difference in distance traversed by one of them is an odd multiple of half wavelengths, then the beams will destructively interfere with each other and no light will be visible (extinction). If the distance is midway between extinction, then reinforcement will occur. Altering the length of one of the light paths will produce a series of alternating extinctions and reinforcements. By counting the number of alternations while moving the mirror a measured distance, a value for the wavelength of light can be computed. The technique of spliting the light beam by a partially reflecting surface, known as amplitude division, was invented in the 1880s by the American physicist Albert Michelson.[80]

Interference is commonly observed in nature. A good example is "Newton's rings," seen when light strikes a thin layer of oil floating on water. Since these are commonly seen in daylight, the white light produces colored rings. Interference patterns can be used to measure other very close interfaces, such as when comparing one optical flat to another. The pattern of the alternating light/dark bands (fringes) gives an indication of how coincident the surfaces are. This technique is important in the commercial manufacture of precision optics.

W. & L. E. GURLEY'S INSTRUMENT MANUFACTORY

ESTABLISHED 1845
INCORPORATED 1900

W. & L. E. GURLEY

MANUFACTURERS OF
CIVIL ENGINEERS' AND
SURVEYORS' INSTRUMENTS
514 FULTON STREET

ORDER NO.
TROY, N. Y., U. S. A., Oct 30 1908
REQ. NO.
SOLD TO Dartmouth College
PACKAGES BOXES 1 Dept. of Physics, Wilder Laboratory
SHIPPED BY Ex. Hanover N.H.

1	No 7024 Interferometer	155 00		
1	Special Fabry & Perot Attachment for #7024	35 00		
1	No 5967 Spirit Level	4 50		
1	No 6005 Spherometer Holder	2 50		
1	No 5996 Spherometer Nut and holder	10 00		
		207 00		
	Allowance	20 70		
		186 30		
	Packing and box at cost	70	187 00	
	By Prof. G. F. Hull.			

NO CASH DISCOUNT ALLOWED
GOODS PACKED AND DELIVERED
F.O.B. TROY, N.Y.

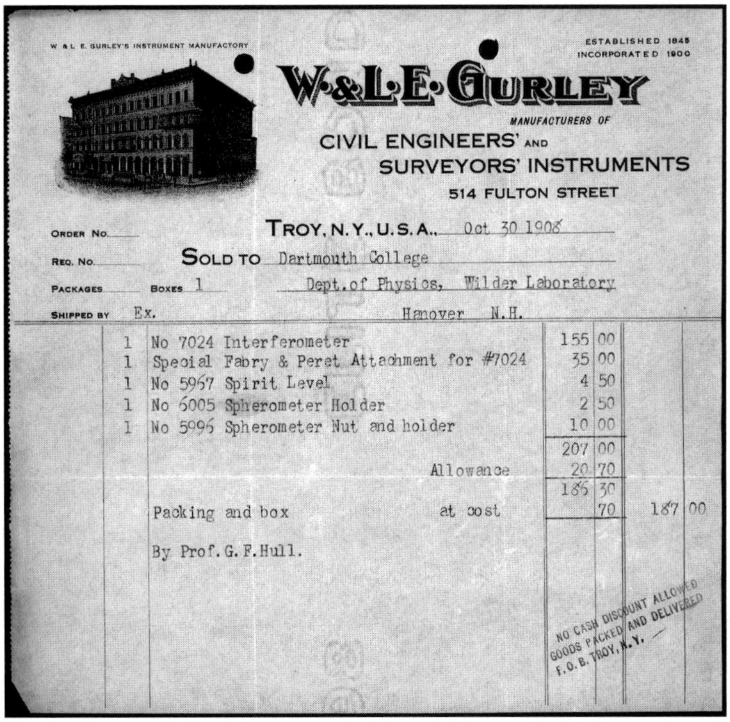

An instrument such as the one shown here was commonly used in student laboratories. A drawing of this Gurley instrument appears in the popular textbook, *University Physics*, authored by MIT professors Francis W. Sears and Mark Zemansky. The first two editions of this text, published before Sears moved to Dartmouth in 1957, do not discuss the interferometer but the third (1964) through the seventh (1987) edition consistently include a drawing of this instrument. The presence of this interferometer at Dartmouth appears to have influenced the presentation of physics to the more than one million purchasers of this textbook.[81]

144 *W. & L. E. GURLEY, TROY, NEW YORK*

No. 7022 This instrument has a heavy nickeled cast iron bed 25 cm. long, giving the carriage a 12 cm. range of motion. The screw is fitted with a worm wheel of 100 teeth and divided on its face in 100 parts. The worm, having an index head divided in 50 parts, may readily be disengaged, allowing the rapid movement of the carriage. The mirrors are 25 cm. square. The fixed mirror and the compensator are provided with a delicate adjustment. The head of the compensator adjusting screw is divided in 100 parts. The instrument is furnished with japanned legs. Price $150.00

No. 7023 Similar to No. 7022, but provided with a counter for automatically counting the number of turns of the screw. This counter may be easily disengaged and used for counting the fringes. Price $155.00

No. 7024 Similar to No. 7022, but with compensator and its adjustment mounted on a torsion rod with a tangent screw-allowing a rotation of 15 degrees Price $160.00

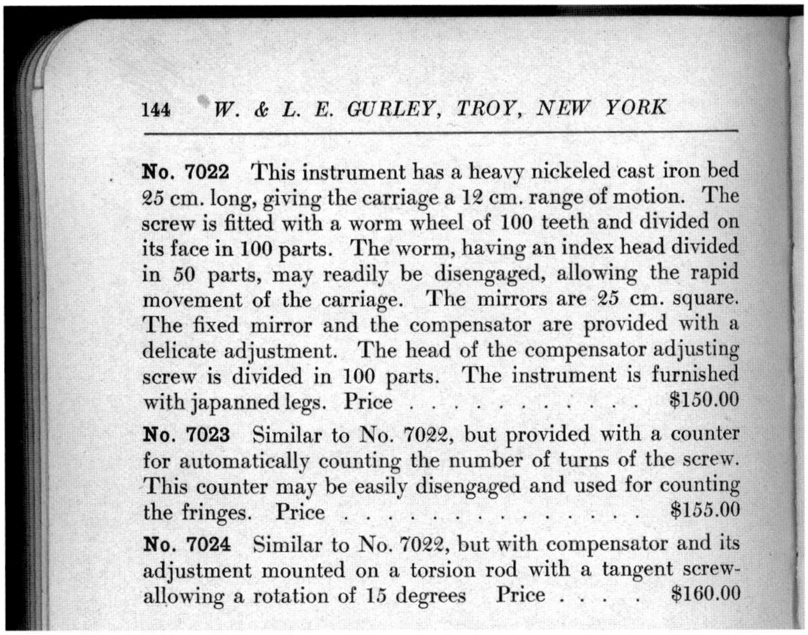

We can put the cost of instruments into perspective by comparing it to other contemporary prices. For this case, in 1908, the basic interferomter cost $155. On May 24, 1907, the trustees voted the following annual salary scales for Dartmouth College faculty at time of their appointment: full professor, $2400; assistant professor, $1600; instructor, $1200.[82]

AIR PUMP

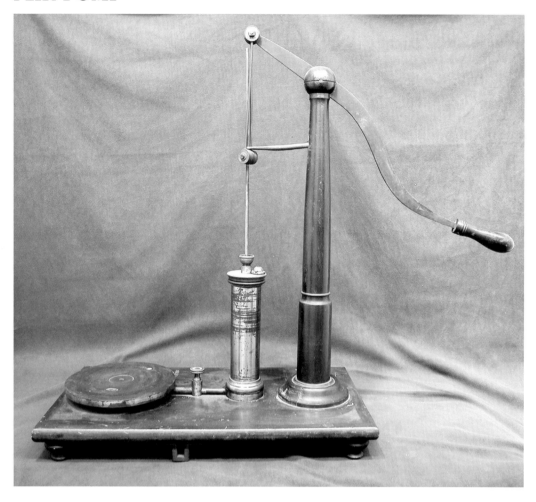

Unsigned (Probably N.B. Chamberlain & Sons)
Boston, Massachusetts
Mid-nineteenth century
Height of handle pivot from base: 21 inches; Base: 12 x 22 inches   2002.1.35231

Hand-operated air pumps were used to obtain a partial vacuum. A small tube at the top of the cylinder can be used to pump air into a condenser (see entry, page 125).

Originally designed in the seventeenth century as research apparatus, air pumps were later frequently used to demonstrate pneumatic phenomena. Common demonstrations included showing the necessity of air to propagate sound and corroborating Galileo's famous "experiment" in which he stated that the rate of fall of an object is independent of its weight. In J.L. Comstock's popular text on natural philosophy (over 200 editions between 1830 and 1860) students were instructed in the use of this model air pump for a variety of experiments, including: "If a withered apple be placed under the receiver, and air is exhausted, the apple will swell and become plump, in consequence of the expansion of the air which it contains within its skin."[83]

By the 1850s, Boston had become a thriving center for the production of scientific instruments. On Washington Street one could find the shops of N.B. Chamberlain, E.S. Ritchie & Sons and Daniel Davis, Jr. By 1857, Alvan Clark & Sons were located across the Charles River in Cambridgeport.

122

Chamberlain's 1844 catalogue shows a pump very similar to this one, and another with two barrels for easier exhaustion of air "used by ladies in our female seminaries, and...reported as working easy and producing a very perfect vacuum." The testimonials in this catalogue for Chamberlain's "largest and most expensive Air Pumps" include "Professor I. Young, Dartmouth College, Hanover, N.H." A very similar illustration appears in Ritchie's 1860 catalogue. According to historian Deborah Warner, a credit report of 1858 indicates that Ritchie paid Chamberlain the sum of $20,000 for permission to "associate in bus[iness]with him that he might learn the trade." A pump based on this design, called a "Ritchie air pump," was still being sold in 1921 by L.E. Knott.[84]

An eighteenth-century copperplate engraving showing pneumatic demonstrations

MANOMETER

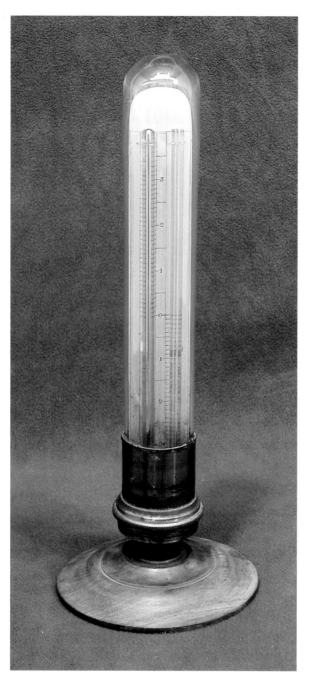

JAS GREEN
New York, New York
About 1850
Height: 31.5 cm   2002.1.35280

The thick glass containing this bent-tube mercury manometer is set into a thin-wall rolled brass collar, about 4 cm in diameter. A cement substance, presumably acting as a seal, is visible. The collar, in turn, is set into a heavy brass base with a threaded orifice at the bottom. The base is knurled to facilitate screwing and unscrewing the manometer. The manometer is screwed into a turned wooden base 12.3 cm in diameter, a support when the manometer is not in use.

Mercury fills the entire left part of the bent manometer tube, which implies that this instrument was used for measuring vacuum, not pressure. We have been unable to fit this manometer to any of the air pumps in the Dartmouth collection, suggesting either that an adapter is missing or that we no longer have the pump to which it was fitted originally.

The manometer was manufactured by James Green, one of the earliest makers of meteorological instruments in the United States. Green was born in London and migrated to the United States in his youth. He apprenticed with an instrument maker and established in 1832 his own "philosophical instruments" firm, making his reputation by a novel barometer that was adopted as a standard by the Smithsonian Institution for their national weather stations. His nephew, Henry J. Green, joined him in 1864. The firm, under the nephew's name (see anemometer entry, page 69), would become a prolific manufacturer of meteorological instruments in the twentieth century.

Condensing chamber

N.B. CHAMBERLAIN & SONS
Boston, Massachusetts
About 1840
Height: 45 cm; Diameter: 18 cm   2002.1.35283

Made of thick-walled glass, this vessel was designed for pneumatic demonstrations involving pressure, rather than vacuum. The manometer at the right reads the internal pressure and the petcock isolates the chamber from the pump. Chamberlain's 1844 catalogue lists this chamber for $10.[85]

AIR PUMP WITH FOUNTAIN IN A VACUUM

"A new air pump, costing about $300, has been added to the apparatus of Professor Young's department," announced the Dartmouth student newspaper in 1869. This large, imposing rotary air pump is evocative of the machine age. The attractively finished mahogany along with the brass mechanism and prominent flywheel make this one of the more aesthetically pleasing instruments in the collection. The inertia of the flywheel aids in operating the pump. The flywheel has a round groove molded into its periphery to accommodate a round leather belt, permitting the pump to be operated by an engine. Rotary motion is transmitted to the piston rod by the crankshaft (see illustration, next page) running between two split bearings and is converted to reciprocating motion by restricting its movement within two parallel rails. The cylinder, about 18 cm long, is marked *patented Nov 19, 1867* (U.S. patent 71,218). Assuming an inside diameter of 9 cm (OD is 9.5 cm) with a measured 13-cm stroke gives the pump about a 900 cc displacement. According to a list of Dartmouth's Appleton apparatus, this pump could exhaust to a pressure of 0.5 mm Hg.[86]

The fountain in a vacuum is attached to the air pump by two petcocks. The air pump has an integral petcock with a

E.S. RITCHIE & SONS
Boston, Massachusetts
1869
Pump: 130 x 56 x 55 cm
Fountain length: 76 cm

2002.1.35288
2002.1.35289

126

cast brass handle that indicates if it is open. The fountain has its own petcock, which may be closed when the fountain is evacuated, permitting removal of the fountain from the pump while retaining its vacuum. When the lower end of the evacuated fountain is immersed in water and the valve is then opened, the pressure of the atmosphere forces water up through the jet pipe (see illustration, previous page).

According to Ritchie's catalogue, "The rapidity of action, the ease of working, and the very high degree of rarefaction obtained by this pump, and the comparatively small space it occupies render this a very valuable instrument. It has been in use and highly approved in many of the principal colleges in the country." This rotary air pump, listed in Ritchie's 1878 catalogue at $250, was one of the most expensive items he sold. The fountain cost an additional $6.50.[87]

Ritchie & Sons made most of their own apparatus, but also sold items produced in Europe. The preface to their 1878 catalogue explained: "Our manufactory is the largest in the United States, and is furnished

with every facility obtained by the best machinery; there is, however, a class of apparatus (made chiefly by hand labor) which can be provided in Europe at lower prices than we make them of equal quality, and we have received from the first makers of London, Paris and other cities agencies for productions, which we will import to the order of Colleges and Universities free of duty."[88]

128

HOUSING LATE-19th-CENTURY PHYSICS

Built in 1840 to house the college libraries and other collections, Reed Hall also served as home to the Department of Physics until 1901. These albumen prints, taken by Hanover photographer, H.O. Bly, mounted on mottled grey board, show the physics lecture hall in Reed. In the top photograph, a vented projection lantern can be seen at the rear of the hall. An assembly of Dartmouth's pneumatic apparatus, placed at the front of the hall, is featured in the lower photograph. Bly made similar photographs of Dartmouth's optical, electrical, thermal and acoustical apparatus.

Bly's photographs were exhibited at the Philadelphia Centennial Exposition in 1876, along with drawings of the college's grounds and buildings, and examples of student work. In the latter decades of the nineteenth century, American colleges and universities began competing for the best-equipped laboratories and lecture halls for science education. Hence, the College's interest in proudly displaying its philosophical apparatus at the Exhibition.[89]

Physical Lecture Room.

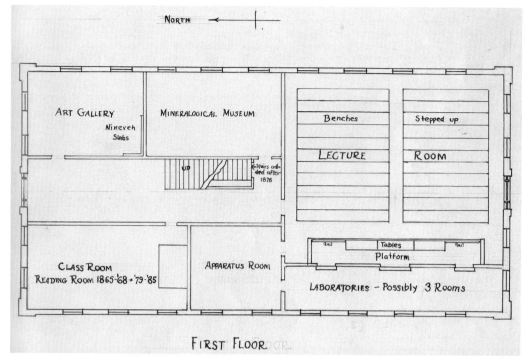

NORTH ←

ART GALLERY
Nineveh
Slabs

MINERALOGICAL MUSEUM

UP

Stairs ad-
ded after
1876

Benches

Stepped up

LECTURE   ROOM

CLASS ROOM
READING ROOM 1865-'68 + '79-'85

APPARATUS ROOM

Rail   Tables   Rail
Platform

LABORATORIES - Possibly 3 Rooms

FIRST FLOOR

The original floor plans of Reed Hall are no longer extant. The above is a reconstruction from 1932 based on interviews with people who remembered the original. These are the first spaces on campus built specifically for the practice of science. From 1841 until 1899 lectures on natural philosophy occurred in the lecture room seen at the upper right and in the Bly photographs. Significantly, the art gallery and mineralogical museum were housed on the same floor. With the College's libraries on the second floor, Reed Hall housed Dartmouth's collections.

Reed Hall still stands with its exterior unchanged but having undergone a succession of uses. In 1899 physics moved into its own building, Wilder Laboratory (below). In 1901, physics professor Ernest Fox Nichols described the new laboratory and its furnishings for readers of *The Physical Review*.[90]

130

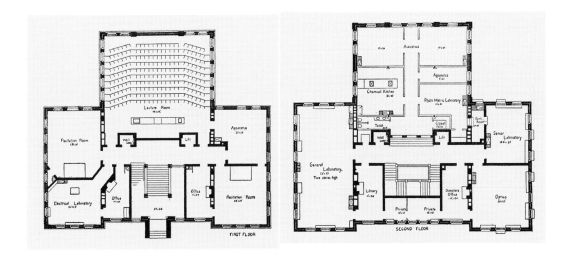

The floor plans Nichols included in his article illustrate dramatically the profound changes that physics underwent in the previous fifty years. Gone are the museum rooms and general libraries. Instead, we see lecture rooms and purpose-built laboratories. Not shown here are the other floors which contain offices, photographic darkroom space and, in the basement, a magneto room with a nearby battery room. This is a building that reflects the maturity of a discipline by recognizing the specific requirements of its subspecialties. The Shattuck Observatory is on a knoll a short walking distance behind the Wilder Laboratory.

CENCO HYVAC ROTARY VACUUM PUMP

CENTRAL SCIENTIFIC CO.
Chicago, Illinois
About 1940
22 x 45 x 16 cm

2002.1.35281

Over the course of the nineteenth century the term "air pump" was replaced by "vacuum pump." These latter pumps were designed for vacuum only, unlike the earlier air pumps that could also be used as compressors. Pumps used specifically to compress air were known as condensers.

Rotary vacuum pumps utilize vanes, or as in this CENCO Hyvac, a rotating eccentric driven by a fractional horsepower electric motor. All the internal parts run in a special oil of low vapor pressure. These pumps were able to produce vacuum, according to the catalogue, of 10^{-3} torr and had a reasonably large pumping volume. When used as a backing pump coupled to a Langmuir-type diffusion pump originally utilizing mercury, ultra-high vacuum can be achieved. The rubber tube attached to the inlet side of the pump has a very thick wall to prevent collapse under vacuum.

Pumps such as the CENCO Hyvac, noted for reliability, served both industry and the laboratory and remained in production for a long time. First introduced in 1921, the Hyvac in 1936 listed for $80.[91]

Founded in 1900, the Central Scientific Company became one of the leading suppliers of teaching and research instruments in the United States, as suggested by the sheer size of their 1936 catalogue (1636 pages). During the post-Sputnik era, sales of laboratory apparatus grew enormously. Central Scientific products from those years still line shelves in many science departments across the country. By the 1950s, the center of the U.S. scientific instrument trade shifted from Boston to Chicago, where firms such as Central Scientific, Welch Scientific, E.H. Sargent Company and The Scientific Shop were based.

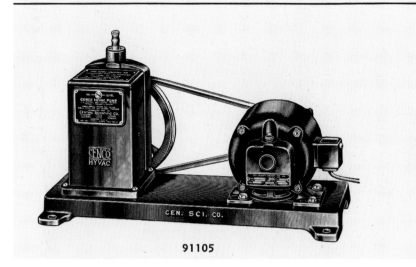

91105

CENCO-HYVAC

91100 **VACUUM PUMP, Cenco-Hyvac, Unmounted, (Patent Nos. 1,719,747; 1,845,216; 1,890,572, and 1,890,614),** an invaluable, high vacuum, mechanical pump of relatively small dimensions and extremely permanent characteristics, which has become the universal favorite of scientists throughout the world for the exhaustion of vessels, chambers, and other vacuum systems in analytical procedures and research investigations.

The pumping mechanism consists of two series-connected, rotating-eccentric pumps joined together to form a single unit. This pumping unit is completely immersed in high vacuum sealing oil contained in a rectangular cast-metal housing which is provided with a trap to prevent the entrance of foreign materials into the pumping mechanism. This trap also serves to retain oil that otherwise might back up into the apparatus under exhaustion when the pump is stopped without equalizing the air pressure within and without the system.

The Cenco-Hyvac pump is guaranteed to attain an ultimate pressure in connection with a leak-proof system of 0.3 micron of mercury pressure. At its normal operating speed of 350 rpm, it has a free air displacement of 10 liters per minute.

The unmounted Cenco-Hyvac pump is furnished with a sub-base to allow the grooved pulley to clear the surface upon which the pump rests. An electric motor of ⅛ hp with grooved pulley is sufficient for operation of the pump. A No. 91130 Molded Rubber V-Belt is recommended for trouble-free operation. Complete with sufficient high vacuum pump oil to fill the pump and directions for care and operation of the pump, but **without** motor base or electric motor.
....**Each 50.00**

91105 **VACUUM PUMP, Cenco-Hyvac, Motor Driven,** compact combinations of a No. 91100 Cenco-Hyvac Vacuum Pump with an electric motor on a common substantial iron base. The entire units are finished in non-chipping black japan. Complete with base, electric motor with connecting cord and separable attachment plug, supply of oil, and directions for care and operation of the pump.

Specifications

Dimensions, over all:		Amount of oil required, qt	1
Height, inches	10½	Net weight, lbs.	54
Width, inches	9	Shipping weight, lbs.	80
Length, inches	18½		

No.	A	B	C	D	E	F	G	H
For current	A.C. 60 cy		D.C.		A.C. 25 cy		A.C. 50 cy	
For volts	110	220	110	220	110	220	110	220
Motor horsepower	⅛	⅛	⅛	⅛	1/12	1/12	⅛	⅛
Each	**75.00**	**77.50**	**80.00**	**82.50**	**80.00**	**80.00**	**80.00**	**80.00**

J136

Central Scientific catalogue, 1936

THE DARTMOUTH ACCELERATOR

Wilder Laboratory, 1962

In 1955, news of New Hampshire's first "nuclear accelerator" circulated in national media. Designed by Leonard Rieser, an assistant professor of physics and Dartmouth's only Los Alamos veteran, and several other members of the Physics Department, this "atom smasher" attracted considerable local attention in the early years of the Cold War. By accelerating deuterium ions through 135,000 volts and smashing them into deuterium targets, the accelerator produced energetic neutrons through nuclear fusion. These neutrons in turn were used to bombard other materials, creating radioactive isotopes whose decay could then be studied. "Construction of the accelerator...is a milestone in physics at Dartmouth," according to *Industrial Science and Engineering*. "It is one of only three in the nation specifically designed to be used by undergraduates in laboratory work sessions." Although several master's theses were written about the apparatus, no publishable research resulted from the Dartmouth accelerator. It remained a teaching tool through the mid-1960s.[92]

The accelerator tube, shaped like a cross at the center of the equipment, required a vacuum of 10^{-5} torr, far beyond the capability of mechanical vacuum pumps. To reach such pressures, the system was evacuated by a Cenco Hypervac (forepump) with a diffusion pump between the Hypervac and the accelerator tube. The diffusion pump can be seen attached to the bottom arm of the accelerator tube.

Diffusion pumps utilize a liquid (at first mercury, later petroleum-based or synthetic oils) that is heated. The ensuing vapor is directed through a nozzle in the center of an aperture. As the vapor molecules pass

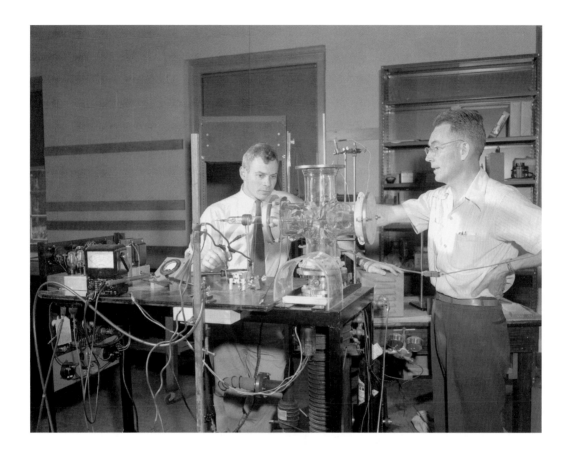

through, they "drag" residual air molecules along, reducing the pressure on the "upstream" side and increasing it on the "downstream" side, creating sufficient pressure for the mechanical pump to evacuate. The mixture of vapor/air is cooled to condense the vapor and the liquid returns to the boiler and recirculates.

The diffusion pump (Model MC 273-01) is seen in the next entry. The outer coils provide coolant (tap or chilled water) that causes the pumping vapor to condense. The top orifice opens to the high vacuum chamber; the mechanical forepump is attached to the lower right orifice.

The illustration above shows physics faculty members Leonard M. Rieser (left) and Willis M. Rayton in 1955 beginning to assemble the apparatus for the accelerator. The accelerator tube, with attached diffusion pump, is visible between the two men. The photograph on the preceding page shows the completely assembled accelerator in 1962.

Built in-house, this instrument demonstrates the usual transparency of such constructs.

DIFFUSION PUMP

CONSOLIDATED VACUUM CORPORATION
Rochester, New York
About 1955
Length: 82 cm   2002.1.35304

KOENIG SOUND ANALYZER

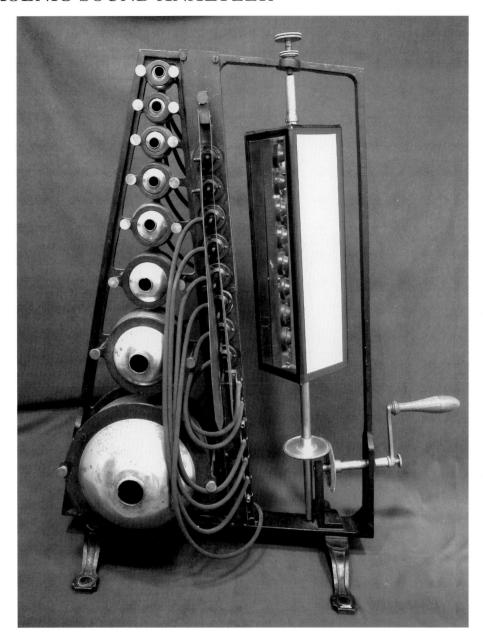

RK. (Rudolph Koenig)
Paris, France
About 1880
Height: 88 cm   2002.1.34112

    The sound analyzer visually displays sound waves using a vibrating flame. It was invented in the early 1860s and was based on Hermann Helmholtz's theory of timbre or *Klangfarbe* (sound color). The analyzer separates the simple, or elemental, tones from the more complex vibrations of a musical instrument or human voice.

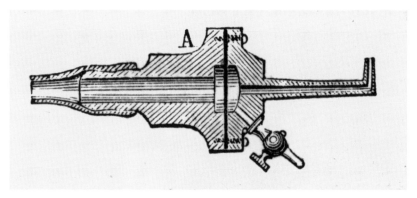

Manometric capsule

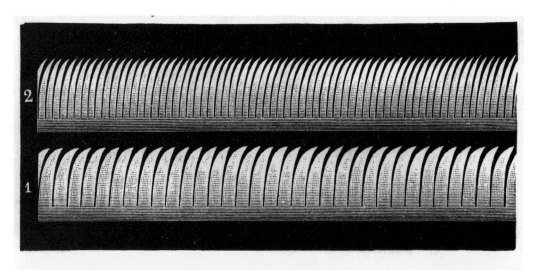

Fig 22. Comparaison des vibrations de deux tuyaux d'orgue par les flammes manométriques.

Spherical brass resonators at the left pick up the elemental tones; the vibrations travel down a rubber tube into the back of a capsule with a membrane. The membrane vibrates, causing gas in the capsule to vibrate. This vibration in the gas produces an imperceptible flicker in a lit gas flame. Rotating mirrors at the right make this flicker visible through the phenomenon of persistence of vision, creating a characteristic wavelike pattern to the flame. Eight different frequencies can be analyzed by this device.

This instrument was enormously popular for demonstration purposes in the nineteenth century and was most likely used by Dartmouth Professor Charles F. Emerson in his Reed Hall lecture demonstrations. The shape of the analyzer's frame resembles Helmholtz's diagrams of the tapering basilar membrane of the ear, and the resonators reveal the mechanical workings of sound analysis. This transparent nature served an important function in the classroom and may explain the widespread appeal of the analyzer.

Rudolph Koenig was the leading acoustical instrument maker of the second half of the nineteenth century. Originally trained as a violin maker under the celebrated J.B. Vuillaume, he later turned his skills toward making acoustical apparatus for scientists. His apartment on Quai d'Anjou near the Notre Dame cathedral combined living quarters, workshop, commercial space and private laboratory. Koenig was best known for his inventions such as the analyzer, and for his precision tuning forks, examples of which survive in the King Collection.[93]

138

This albumen print from the 1876 Bly photographs shows a collection of instruments for demonstrating acoustical and wave phenomena, assembled in the Reed Hall lecture room.

REED PIPE AND RESONATING PIPE

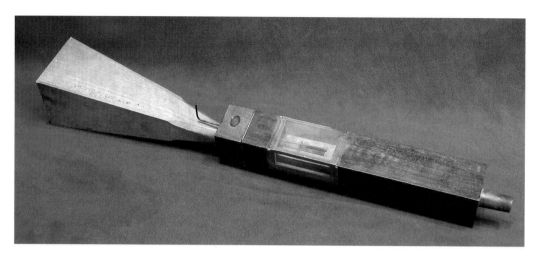

MARLOYE ET CIE
Paris, France
About 1853
Length: Reed pipes, 33 cm; Resonating pipe, 24 cm.

2002.1.34049
2002.1.34050
2002.1.34089

For millennia, vibrating reeds have been used to produce musical sounds. A reed is an elastic substance (originally vegetable) that vibrates when subjected to a stream of air. The frequency vibration depends on a number of factors, including the shape, composition and mass of the reed. These vibrations induce sound waves in the instrument, the form of which determines the timbre of the sound emitted.

There are generally two types of reed pipes. "Beating" reeds (bottom detail, top illustration, next page), found for example on the clarinet and saxophone, have a tongue that completely covers the aperture or cavity beneath the reed. Pulsations of air result as the reed opens or closes (i.e. beats) against the aperture. "Free" reeds (upper detail), on the other hand, oscillate freely within the aperture as air passes around them.

Controversy arose over the relative value of these types of reed pipes in musical instruments. Introduced into European organs in the latter part of the eighteenth century, free reeds were said to reduce the harshness of sound often experienced with beating reeds and to enhance "expressiveness," according to their defenders. By the early nineteenth century, free reeds became popular especially in Germany. By midcentury, the famous French organ maker, Aristide Cavaillé-Coll, also began using them. Hermann Helmholtz in 1863 declared that free reeds had universally replaced beating reeds. However, in 1885, Alexander Ellis asserted that English makers had never used free reeds, and noted that even Cavaillé-Coll had eventually abandoned them. According to Ellis, organ makers complained that free reeds lacked the power of beating reeds.[94]

Small vibrating reed pipes like those pictured here were employed in laboratories for research on the phenomenon of sound as well as in attempts to elucidate the basis of musical instrument performance. The tone of both pipes (next page) is nasal and rather weak when blown using breath. Both are marked UT_2, a frequency roughly equivalent to "middle C" on the piano. As is the case with conventional organ pipes, a tuning wire emerges from the boot of each pipe. The red paper tape framing the windows in the pipes is original, and appears similar to tape frequently found on Marloye reed pipes.

The pipes are accompanied by two large and two small wooden resonating pipes, also made by Marloye. The above picture shows one of the small resonators. The reed pipes are made to fit into Marloye's bellows, visible at the lower left in the Bly photograph of acoustical apparatus (see previous page).

Free reed–upper left; beating reed–lower right

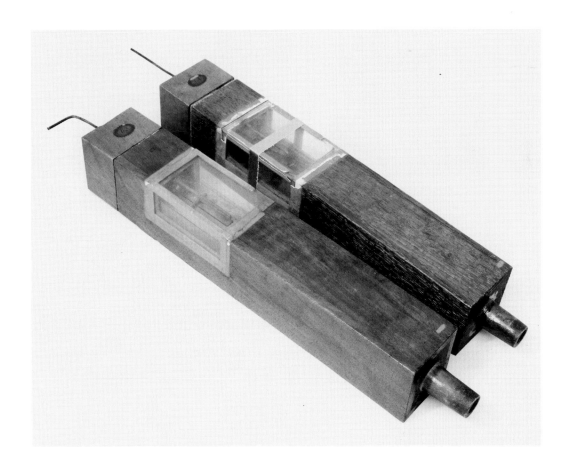

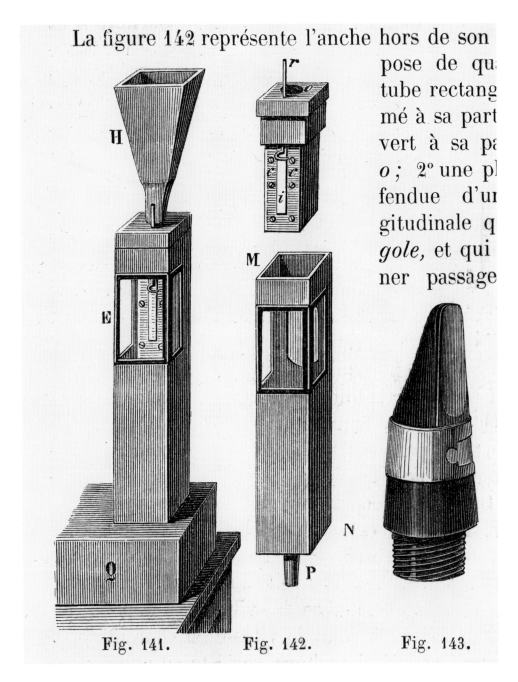

H

r

i

M

E

N

P

Q

Fig. 141.   Fig. 142.   Fig. 143.

These pipes were purchased in 1853 by Ira and Charles Young during their trip to Paris. Marloye sold the beating reed for 18 francs, the free reed for 20.[95]

This illustration from Adolphe Ganot's physics textbook (1856) shows the construction details of the pipes.[96] The front side of the pipe has a glass face that permits one to view the movement of the reed. The opposite side is covered with a leather membrane that one can push with a finger to distort or silence the pipe.

Vibrating plate with lycopodium

MARLOYE ET CIE
Paris, France
About 1853
Stand height: 22 cm; Plate diameter: 35 cm

2002.1.34026

   Stiff metal plates, sprinkled with sand and made to vibrate with a violin bow, were among the best-known acoustical demonstrations in the nineteenth century. As sand gathers at the nodal positions (places of no vibration), beautiful geometric patterns emerge on the plates. Ernst Chladni, amateur musician and pioneer of experimental acoustics, first discovered this effect and after the 1802 publication of his book, *Die Akustik*, such vibrating plates became known as Chladni plates.

   The instrument above includes a resonator tube to reinforce the vibrating segments. In the 1830s, the Parisian maker Marloye had developed an extensive line of instruments for Félix Savart's lectures on acoustics at the Collège de France. Dartmouth's instrument is a variation on the popular Savart bell, in which a brass bell is set into vibration and a resonator placed beside the bell enhances the vibrations. The pasteboard resonator tube was made to be open at the top but ours is crudely closed with additional pasteboard.

   Nineteenth-century experimenters discovered different effects by using a much lighter powder, lycopodium (spores from moss, especially *L. clavatum*), to reveal even more intricate patterns on Chladni's plates. Hans Christian Ørsted, Savart and Michael Faraday claimed to find effects with this powder that Chladni had missed. In his 1851 catalogue, Marloye described the above device as an "apparatus for demonstrating that the rotation of lycopodium on a round plate is due solely the translation of nodal lines around the circle." Rather than marking nodal lines, the light powder forms dust swirls at the antinodes (places of greatest vibration). On the apparatus above, the swirls and nodal lines are supposed to rotate around the circular disk. However, contemporary handbooks noted how difficult it was to obtain such rotational effects.[97] Like all Chladni experiments, working this apparatus requires much tacit skill. One has to produce the fundamental frequency of the plate (the easiest frequency with which to work), tune the resonator accordingly, and also learn to draw out other frequencies at will. The illustration below shows the patterns of lycopodium powder elicited by the vibrations of the plate on our instrument when set into motion by vigorous bowing.

In the nineteenth century, Chladni patterns also captivated the popular scientific imagination. Praising the blissful effects of travel by rail, Oliver Wendell Holmes (who had briefly taught medicine at Dartmouth) in 1862 described "[m]y thoughts shaken up by the vibrations into all sorts of new and pleasing patterns, arranging themselves in curves and nodal points, like the grains of sand in Chladni's famous experiment."[98]

When Charles and Ira Young visited Paris in 1853, one of their first stops was Marloye's atelier near the Pantheon. Charles wrote in his diary: "He is a pleasant old gentleman about 60, a member of the institute & red ribbon wearer, and though he can understand no word of English, written or spoken, he very kindly showed us many interesting experiments upon sound, with his different apparatus."[99]

PROJECTION WAVE MACHINE

MAX KOHL
Chemnitz, Germany
About 1903
Box: 19 x 10 x 20 cm
Stand height: 20 cm
Disks: 20 cm diameter
2002.1.34024

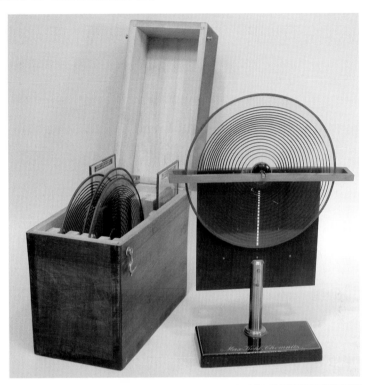

This instrument contains a set of six glass elements, four circular disks and two rectangular plates, with tape strips on their surfaces and a cast iron base with brass frame. The latter is painted black with a golden stripe, a characteristic decorative feature of Kohl instruments. The base is signed *Max Kohl, Chemnitz* in gold.

Supported by the frame, the disks are rotated by hand. A collimated beam of light, directed against the black square behind the disk, passes through the narrow vertical fixed slit and projects on a distant screen the patterns generated by the tape strips on the rotating glass. The rectangular plates slide in brass holders (not visible in picture) horizontally across the slit. According to Kohl's 1911 catalogue, the projected patterns demonstrate various motions of waves in tubes.

Early in the nineteenth century, wave motion was demonstrated using rotating cylinders or moving trains of glass beads. In 1867, the French physicist André Crova described the first device to project wave motion on a screen for large audiences. At the Paris Exposition of 1867, Rudolph Koenig exhibited Crova's wave machine. Kohl's machine seen above copied the French design. His firm's catalogue offered a number of different wave machines. Around 1903, E.F. Nichols and G.F. Hull ordered several Kohl pieces to equip Dartmouth's new Wilder Laboratory.

At the turn of the twentieth century, Kohl was one of the world's largest suppliers to student and research laboratories. In 1888 Kohl employed 16 workmen; in 1900 there were 148, and by 1911 more than 360 workers labored in their factory (right) at Chemnitz, Germany. Kohl's catalogues appeared in both English and German editions.[100]

145

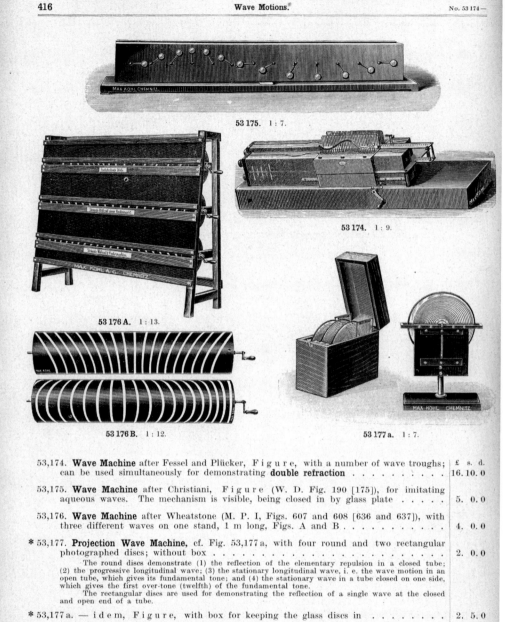

53 175. 1 : 7.

53 174. 1 : 9.

53 176 A. 1 : 13.

53 176 B. 1 : 12.

53 177 a. 1 : 7.

	£ s. d.
53,174. **Wave Machine** after Fessel and Plücker, F i g u r e , with a number of wave troughs; can be used simultaneously for demonstrating **double refraction**	16.10. 0
53,175. **Wave Machine** after Christiani, F i g u r e (W. D. Fig. 190 [175]), for imitating aqueous waves. The mechanism is visible, being closed in by glass plate	5. 0. 0
53,176. **Wave Machine** after Wheatstone (M. P. I, Figs. 607 and 608 [636 and 637]), with three different waves on one stand, 1 m long, Figs. A and B	4, 0. 0
* 53,177. **Projection Wave Machine,** cf. Fig. 53,177 a, with four round and two rectangular photographed discs; without box	2. 0. 0

The round discs demonstrate (1) the reflection of the elementary repulsion in a closed tube; (2) the progressive longitudinal wave; (3) the stationary longitudinal wave, i. e. the wave motion in an open tube, which gives its fundamental tone; and (4) the stationary wave in a tube closed on one side, which gives the first over-tone (twelfth) of the fundamental tone.

The rectangular discs are used for demonstrating the reflection of a single wave at the closed and open end of a tube.

	£ s. d.
* 53,177 a. — i d e m , F i g u r e , with box for keeping the glass discs in	2. 5. 0
* 53,178. **Wave Machine** after Crova, F i g u r e (M. P. II, 1, Fig. 193 [271]; Fr. phys. Techn. I, 2, Fig. 3409 [I, Figs. 508 and 509]), with three discs	2. 0. 0

(1) Stationary longitudinal wave.   (2) Progressive longitudinal wave.   (3) Two longitudinal waves with phase displacement.

* Can be used with the Projection Apparatus.

Cl. 1025, 5541, 1024, 1026³, 1026², 5051.

Max Kohl catalogue, 1911

NAIRNE-STYLE CYLINDER ELECTRIC MACHINE

Unsigned (John Prince?)
Salem, Massachusetts
About 1810
Cylinder: 12 x 18 inches; Overall: 27 x 33 inches

2002.1.34867

Prior to the invention of the voltaic pile (see entry, page 153), static electricity was the only form known and was most easily generated by friction. In the machine above, a horizontal glass cylinder, when rotated, rubs against a leather or silk cushion (missing) to produce electric charge. The vertical glass column, mounted on a sliding board secured by two threaded wooden pegs, once held the cushion in place. The mahogany crank handle and the wooden slider supporting the glass column may be later replacements. To use the machine, an insulated prime conductor (also missing) is placed beside the cylinder opposite the cushion. A metal comb, attached to the conductor with its teeth perpendicular to the cylinder, collects the charge which is then either used directly or transferred to a Leyden jar (see entry, page 151).

Although small cylinder machines had been used since the 1740s, it was the London maker, Edward Nairne, who in 1774 popularized the design by constructing the largest and most powerful cylinder machine to date, with a drum 19 inches in length and 12 inches in diameter. In 1782 Nairne patented a smaller, more compact cylinder machine that became widely used for medical therapy as well as electrical studies.[101]

In the third edition of George Adams' widely read *Essay on Electricity* (1787), a Nairne machine is shown charging a prime conductor connected to a Leyden jar. A spark gap at the top of the jar regulates the strength of the shock being administered to the young girl, who does not look particularly overjoyed by the procedure.

Dartmouth's first electrical machine, described as "small" by a visitor to the College in 1787, had been obtained from England several years earlier. The above cylindrical machine is unsigned, but has long been related by the College's faculty to John Prince, of Salem, Massachusetts, the leading colonial and American maker and supplier of philosophical apparatus to the new nation's colleges, academies and schools. In 1814 Professor Ebenezer Adams journeyed to Salem and purchased from Prince various (unspecified) apparatus totaling $544. In 1816, the trustees "voted that Professor Adams be authorized to borrow eighty-five dollars payable when the board shall be in funds, in order to pay the balance now due to Doctor Prince of Salem for articles furnished for the Philosophical apparatus—and that he also be authorized to exchange any articles of said apparatus for such others as he shall deem more useful and necessary."[102]

During the course of the nineteenth century, Dartmouth faculty developed a story that this machine came from Prince, along

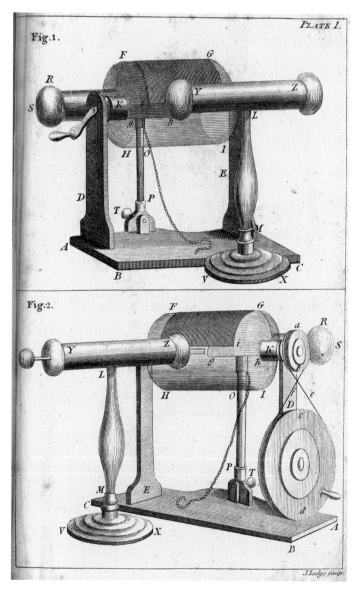

148

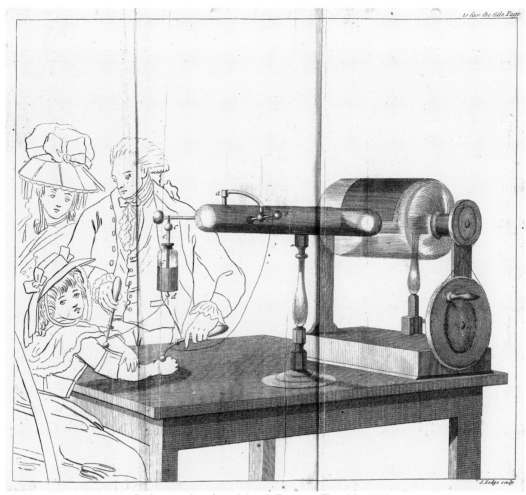

Both engravings from Adams' *Essay on Electricity,* 1787

with several other articles including a 36-Leyden jar battery (see entry Bly photographs, page 163 top). The latter was said to have once belonged to Joseph Priestley, who in turn had acquired it from Benjamin Franklin. No extant evidence supports the Priestley/Franklin connection. On the other hand, Prince is known to have provided or repaired cylinder electric machines for Harvard, Rhode Island College (now Brown), Bowdoin College and the University of Vermont. He procured a similar apparatus for Dartmouth in 1817 (see page 218).[103]

ELECTRIC PLUME; FRANKLIN BELLS

Electric plume (left)
CENTRAL SCIENTIFIC COMPANY
Chicago, Illinois
About 1940
Height: 17 cm   2002.1.34899

Franklin bells (right)
Unsigned
1780-1830
Diameter: 6 cm   2002.1.34915

Early demonstrations of static electricity often involved use of human subjects, both to reveal the phenomena more directly and to entertain. However, a large number of inanimate objects were devised to demonstrate electrical effects. These toy-like items are qualitative detectors of electric charge.

The Franklin bells discharge a Leyden jar. The middle bell, suspended by a non-conducting thread, has a chain hanging from it to make contact with the conducting rod of a Leyden jar (not shown above). The outer surface of the Leyden jar would be placed in contact with the stand supporting the bells and the outermost bells suspended by chains. This arrangement produces a potential difference between the central and outer bells that is equal to that between the inner and outer surfaces of the charged jar. In this situation the two metal clappers suspended by non-conducting threads will begin to strike the bells as they are first attracted to a bell, strike it and pick up its charge and are then repelled from that bell and attracted to the opposite bell. As the bells keep ringing, charge is transferred between the surfaces of the Leyden jar until it is completely discharged. Franklin created the first electrically driven chime.

The bell clappers are marked *SET. FAST.*

The twentieth-century electric plume consists of thin strips of paper that, when charged, repel each other. Central Scientific Company's 1936 catalogue offered another version of this plume for 65¢.[104]

150

LEYDEN BATTERY

Unsigned
About 1840
Wooden box: 20.3 x 47 x 47 cm
Diameter of each jar: 12 cm

2002.1.35421

151

Independently invented around 1745 by Ewald Kleist in Germany and Pieter van Musschenbroek and Andreas Cuneaus in Holland, the Leyden jar stores electrical charge. It consists of a glass jar (functioning as a dielectric), with unconnected metal coatings on its inner and outer surfaces, and a central conductor touching the inner coating and extending out of the jar, held in place by a non-conducting cap. When charged with an electric machine, positive and negative charges accumulate on the respective coatings, establishing a potential between them. When the gap between the outer coating and the conducting rod is narrowed, the jar discharges with a spark and loud snap. The Leyden jar is the earliest form of the condenser, or capacitor. It enabled the creation of more powerful electrical phenomena than did simple electric machines, and thus played a significant role in late-eighteenth- and early-nineteenth-century studies of electricity.

The hollow brass ball (7.7 cm diameter) is bored to take the solid brass rods (8 mm) connecting it to each Leyden jar. The brass rods pass through turned mahogany insulating caps (82 mm diameter) and attach to inner chains, which provide electrical continuity with the inside metal coating of the jar. Hence the inner surfaces of all nine jars are connected electrically. Likewise, a single sheet of foil covers the bottom of the wooden box, connecting the external metal coatings of all the jars. Thus the jars of this battery are connected in parallel. The resultant capacity of such an instrument is equal to the sum of the capacities of the individual jars.

The capacity of early Leyden jars was measured by the volume of the jar. The jars of this battery are each about three quarts. A note inside one of the jars reads: "Recoated jars outside. Repaired box. March 1863. H.F."[105]

152

VOLTAIC PILE

Unsigned
Early nineteenth century
Height: 46 cm

2002.1.34905

The voltaic pile was the first source of uninterrupted flow of electricity. Prior to its invention by Alessandro Volta in 1800, electricity had been generated using static apparatus. Volta's pile, generating electricity by stacking pairs of silver coins and zinc discs separated by cloth soaked with salt water (electrochemical generation), drew immediate worldwide attention. The production of a continuous current provided a new tool for studying electricity.

Our electric pile is held between turned wooden ends separated by brass-ferruled glass rods. A wooden screw compresses the pile to ensure contact between the components. The pile has no terminals for attaching wires; rather a spark gap is provided between brass spheres attached to each end of the pile. This suggests that our piece was intended to demonstrate the ontology of electricity rather than to provide a source of current for other experiments.

Ten-cell Grove battery

W. LADD & CO.
London, England
About 1880
About 21.5 x 55 x 11.5 cm
Exhibited in *Mind and Hand*. MIT Museum, Cambridge, Massachusetts, May 31 2001-May 31 2003. 2002.1.35420

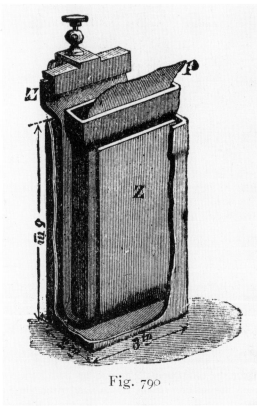

Fig. 790

Wood engraving showing cutaway view of Grove cell

Ladd & Co. made a variety of instruments in the latter part of the nineteenth-century, but this battery was either made in very small numbers or did not survive, for it appears to be rare. All the platinum plates are missing from this specimen.

This Ladd battery contains a series of ten individual Grove battery cells, fitted into an oaken rack. Each cell is made of an impervious material only a few mm thick and non-reactive with the dilute sulphuric acid it must contain. Its composition is not clear, but could be vulcanite. Within this cell is a broad "U"-shaped zinc plate fitted with a connecting screw. Nestled within the zinc "U" of each cell is a porous ceramic pot that bears the impression *Ladd & Co. London*.

The porous pot contained concentrated nitric acid. The thin platinum plate placed in the nitric acid generated a potential between it and the zinc. William Robert Grove invented this arrangement in 1839 by replacing the copper plates of earlier cells with platinum because nitric acid destroys copper.

According to Adolphe Ganot, the Grove design is convenient and powerful, but expensive because of the platinum. In use,

154

the battery releases nitrous fumes, which would have been unpleasant and corrosive. The zinc had to be particularly pure so that bits of metallic impurities did not set up small destructive circuits of their own. The platinum became brittle, requiring heating to re-anneal it. Grove cells were difficult to maintain, requiring frequent cleaning. It was suggested that they be taken apart every night and the electrodes scrubbed in the morning after an overnight soak.

Grove batteries were the favorite source of power in the early days of telegraphy. A single Grove cell might produce more than 10 amps at an emf of 1.8 volts.[106]

The Grove cell was the prototype for several later improvements. Perhaps most notable was that of Bunsen, who by using carbon instead of platinum for the positive electrode reduced the cost without reducing the power of the cell. A single Bunsen cell provides about 12 amperes at 1.8 volts when shorted.[107]

156

COMPOUND MAGNET AND ELECTROTOME

Unsigned (probably Charles G. Page and Daniel Davis, Jr.)
About 1837-38
Base: 7 1/4 x 3 7/8 x 11/16 inches

2002.1.35088

This rough coil, although very simple in construction, may bear witness to a significant moment in the history of electrical instruments. It may be the prototype for the first induction coil manufactured in America.

Starting in 1837, Daniel Davis, Jr. of Boston became the first American to commercially produce apparatus for the burgeoning field of electromagnetism. He worked closely in those years with Charles Grafton Page, an electrical experimenter and inventor living in Salem, Massachusetts, who later would become a well-known U.S. patent examiner in Washington, D.C. In his first catalogue published in November 1838, Davis described what he called "Page's compound magnet and electrotome." In January 1839, Page himself published a brief description and illustration of the device (right). Subsequent editions of Davis's catalogue feature woodcut illustrations quite similar to Page's (see page 159). At first glance, Dartmouth's apparatus appears to be a crude, "homemade" copy of Page's compound magnet and electrotome as produced by Davis.[108]

What Page had created was an induction coil with an automatic (electromagnetic) contact breaker or interrupter. An electric current from a

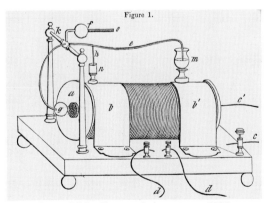

First published image of electrotome, 1839

157

battery passes through a primary coil of thick wire wound on an iron core consisting of several strips of iron wire, creating an electromagnet. As the right end of the pivot arm (labeled an "electrotome" by Page) is attracted to the magnet, its arched end is lifted from a cup of mercury, breaking the circuit. This turns off the magnet and allows the arm to fall again into the mercury, re-energizing the magnet and producing an oscillating magnetic field that induces a current in the secondary coil, comprised of smaller wire wound in many more turns around the primary coil. As the electrotome vibrates, sparks or shocks are produced across the leads of the secondary coil.

The early development of the induction coil in the late 1830s was wildly contentious, as literally dozens of inventors, mechanics, physicians and public lecturers competed to generate ever stronger shocks and sparks (the measure of a coil's "power") by tinkering with designs for the device's core, windings and interrupter. By 1840, the induction coil had become a basic tool for electrical experimenters, popular lecturers and physicians. Although its design remained somewhat fluid, a central feature of the wiring had by that time become fixed, viz., the primary and secondary coils were kept entirely separate. The Page apparatus, as described in 1838 by Davis and then Page, provided separate terminals for the primary and secondary coils. Likewise, all specimens known to us of Page's compound electromagnet and electrotome, made by Davis, are wired with separate primary and secondary coils.[109]

The copper wire used to wind the primary and secondary coils on the Dartmouth apparatus is of the same size as that used on the Davis coils (0.060 and 0.025 inches, respectively). And although the Dartmouth coil has been painted black, small patches of green, the characteristic color used by Davis to paint his coils, appear at the ends of its primary coil where the black paint has chipped away. Such features link the Dartmouth coil to Davis's workshop.

However, the Dartmouth device also differs from those depicted and described by Page and Davis in 1838. The Dartmouth apparatus is wired with its secondary coil in series with the primary coil (an arrangement later denoted as an autotransformer). The profile of the mercury cups on the Dartmouth coil is a simple trapezoid, unlike the more articulated design of such cups characteristically found on Davis coils. And the Dartmouth electrotome lacks the top counterweight and the soft iron cylinder or sphere at the end nearest the core of the coil. Interestingly, each of these unique features can be linked to Page's experiments before 1838.

Page's first published experiments on induction (dated May 1836) feature a spiral of copper ribbon, wired as an autotransformer. His first reported experiments with coils made of iron cores, helically wrapped

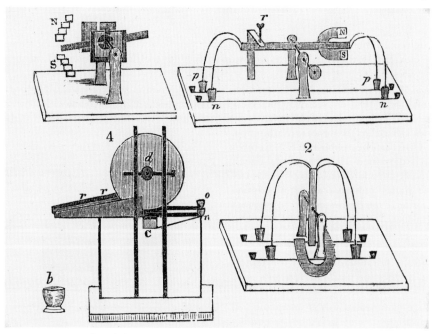

Page's interrupter with trapezoidal mercury cups, 1837

with copper wire (dated June 1838), also combine the secondary and primary coil (an arrangement first employed by the Irish physicist, Nicholas Callan, in August 1836). In this same paper, Page also describes another helix with its primary and secondary coils wired separately (Callan had made this change by September 1837). That Page (and Davis) might have wired a coil like Dartmouth's in 1837 or early 1838 seems quite plausible.

Furthermore, Page's earliest published diagrams of an electrotome interrupter (dated April 1837) show trapezoidal mercury cups (see illustration previous page). Similarly shaped cups appear in an illustration of a pole changer (undated, prior to June 1838) invented by Page, and in an illustration of a different type of electrotome (undated, prior to November 1838). None of Page's subsequent publications depict trapezoidal mercury cups; neither do they appear in Davis catalogues or apparatus made by his shop.

Finally, the Davis catalogue of 1838 offers a quite detailed description of the rocking electrotome on Page's coil. The end to be attracted by the electromagnet of the primary coil, however, is referred to simply as the "extremity of the wire." Page in his article depicted an iron sphere attached to the end of the wire but added that a cylinder of iron would work better. The Smithsonian coil, made by Davis in 1838, has a cylinder at the end of the wire. In his 1867 history of the induction coil, polemically crafted to defend the priority of his inventions, Page describes his experiments on the design on the rocking electrotome. The Dartmouth coil, with its unadorned electrotome consisting of a soft iron wire soldered to a copper wire for making contact between the mercury cups, might thus be a prototype, constructed by Page and Davis in 1837-38 as they struggled to perfect this early form of the induction coil.[110]

An undated inventory of Dartmouth's philosophical apparatus, probably prepared during the 1870s, lists "Page's apparatus for shocks with mercury break," undoubtedly referring to the induction coil. The device is not included in an 1862 inventory of the apparatus, which might imply that Dartmouth acquired the coil sometime after Page died in 1868.[111]

The induction coil, exploiting Faraday's 1831 discovery of electromagnetic induction, is the antecedent of the transformer, the device for stepping voltage up or down that would play a central role in the rise of the electrical industry by the end of the nineteenth century. The high voltages provided by the induction coil would power the cathode ray tubes of the second half of the nineteenth century and thereby provide empirical foundations for the new physics of spectroscopy, X-rays, radio waves, and the elementary particles.

Daniel Davis, Jr., manual, 1842

INDUCTION COIL

APPS
London, England
1871
10 x 26 x 53 cm
Serial number: 796

2002.1.35384

After 1850, instrument makers in Europe and America began to manufacture ever larger induction coils. Working largely by trial and error, makers as well as amateurs explored a host of innovations to increase the spark-producing "power" of the apparatus. Since the voltage induced across the secondary coil is proportional to the ratio of turns between the secondary and primary coils, the output voltage was increased by winding ever more turns on the secondary. Cores consisting of many small wires of soft iron were found to increase spark lengths. By adding a commutator that enabled the direction of current through the primary coil to be periodically reversed, a permanent magnetic field (which would degrade the performance of the coil) was prevented from developing in the core. To prevent arcing and to quench the changing magnetic fields more abruptly, a condenser (capacitor) was placed in parallel with the interrupter.

Perhaps the greatest hindrance to increasing the spark length (or voltage output) of midcentury induction coils was the insulation and isolation of the secondary coil. In the 1850-60s, copper wire usually was wrapped with a layer of silk or cotton thread and then soaked in

paraffin or shellac to create insulation. If neighboring loops of the secondary wire short together, the number of effective turns of the coil is reduced and the coil's performance suffers. Indeed, the life of an induction coil was often quite short, due to failed insulation on the wire of the secondary coil.

Initially, coil makers simply wrapped the secondary wire in helical fashion from one end of the core to the other, occasionally inserting a sheet of insulating gutta percha or paraffined paper between successive layers. In this geometry, however, neighboring layers of wire might be separated by hundreds of loops, creating a potential great enough to breach the insulation and compromise the coil. In 1852, the British inventors Edward and Charles Bright proposed winding the secondary in a series of flat toroidal sections, each separated by a strongly insulating disk. In this geometry, adjacent wires are separated by only a few loops, reducing the potential between them and thereby protecting the insulation on the wires.

By combining many of these innovations, the Parisian maker, Heinrich D. Ruhmkorff, succeeded by the mid-1850s in producing reliable, powerful coils, and on the continent the device became known as the "Ruhmkorff coil" even as American makers such as E.S. Ritchie complained bitterly about the eponym.

Amateurs could find detailed instructions for winding coils in magazines such as *The English Mechanic and Mirror of Science* or *Scientific American*. The schematic diagram on the next page illustrates the basic electric circuitry of the midcentury induction coil. Current passes from the battery (B) through the commutator (C) to the interruptor (Ab), which is bridged by the condenser (CO). The high-voltage output is at TT'.[112]

By 1871, when Charles A. Young acquired the above device, the most powerful coils were being made by Ritchie & Sons of Boston and Alfred Apps of London. The previous year, Young had publicized his purchase of a $700 Ritchie coil 30 inches long and 8 inches in diameter. With a secondary coil containing 35 miles of wire, the coil was, Young claimed, "the most powerful in the country," capable of producing sparks 20 inches in length. This coil, no longer extant in the King Collection (but visible in the entry "Electrical apparatus on display," page 163, lower illustration), represented a major acquisition for Dartmouth's collection of philosophical apparatus.[113]

During his visits to London in 1870-71, Young went to Apps' workshop several times and purchased for £14 14s (roughly $80) a "patent 4 inch coil," apparently the device depicted above (Apps rated his coils by the length of sparks they could generate). Several years earlier, Apps had exhibited a massive, 1500 pound coil at the London Polytechnic that could produce 29-inch sparks from the 150 miles of wire in its secondary coil. By 1877, Apps would construct the world's largest induction coil to date, able to generate 42-inch

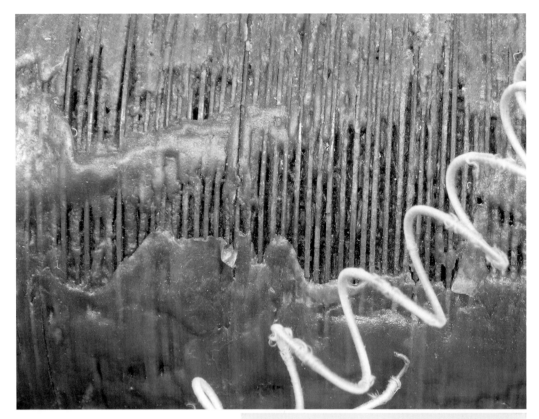

sparks across a secondary coil containing 280 miles of wire! At Dartmouth, the Apps coil produced 6-inch sparks.[114]

Apps generally encased the outside of his coils in leather or cloth. On the Dartmouth coil, this cover is missing and the outer surface is wax. Under the wax we can see about 100 flat toroidal windings, separated by thin discs of ebonite. Individual wires of the secondary coil cannot be seen. If,

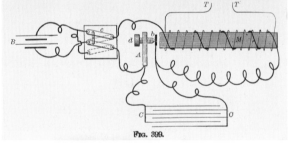

FIG. 399.

however, Apps used the same wire (0.014 inches diameter) for this coil as he did for the London Polytechnic coil, then the Dartmouth coil's secondary winding would be about five and one-quarter miles long. The primary winding is enclosed in an ebonite cylinder seen passing through the secondary. Thumb screws on the interruptor enable the rate of break to be adjusted. The commutator, missing from Dartmouth's coil, has been replaced by a length of wire. The capacitor also has been removed. This coil was used in 1896 to produce the first X-rays at Dartmouth (see entry, page 171).

Apparently, Apps numbered his induction coils sequentially. The Dartmouth coil has No. 796 stamped inside the wooden base and engraved on the irovy nameplate; Apps' massive 1877 coil is numbered No. 847. As noted by Allen King, the Dartmouth nameplate erroneously gives Apps' street address as 133 Strand (rather than 433 Strand).[115]

In addition to astounding public audiences with their giant sparks and to providing high voltages for physics laboratories, induction coils increasingly would find commercial application. Technologies ranging from electric lighting, internal combustion engines, and induction motors to detonation, wireless telegraphy and medical uses all would exploit the physics of the induction coil. In an 1869 advertisement, Apps even extolled their promise for electrochemical applications such as commercial bleaching of cotton and decolorizing of sugar.

162

ELECTRICAL APPARATUS ON DISPLAY

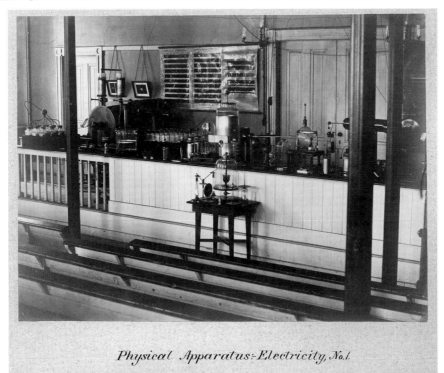

Physical Apparatus=Electricity, No.1.

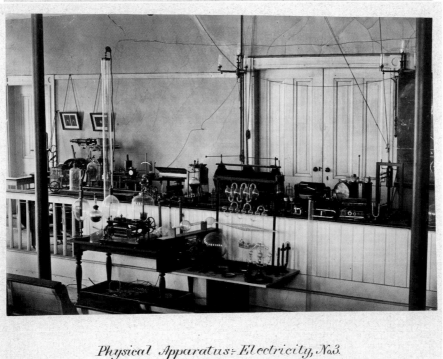

Physical Apparatus=Electricity, No.3.

Bly photographs for the Centennial Exposition, Philadelphia, 1876

GEISSLER TUBES

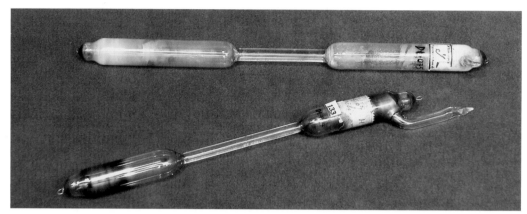

Franz Müller (top) marked *J*[od] (Iodine)
Bonn, Germany
1906-7
Length: 23 cm   2002.1.34721

Unsigned tube marked *Helium* (probably Franz Müller)
1906-7
Length: 20 cm   2002.1.35427

Johann Heinrich Wilhelm Geissler was a glassblower, following the family tradition, and a physicist at the University of Bonn. He produced glass tubes, filled with highly rarified gases, with electrodes embedded in each end. Gases at normal atmospheric pressure do not conduct electricity very well. However at lower pressures, when a high voltage is applied to the electrodes, the gas ionizes and acts as a plasma, glowing with color characteristic of its composition. Geissler's tubes demonstrated these phenomena well, and in 1854 he opened a shop where he made and sold them. He also devised a vacuum pump that employed falling mercury to evacuate the tubes.

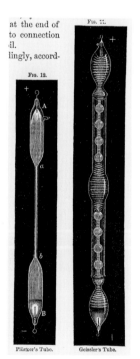

at the end of
to connection
il.
lingly, accord-

Because the glowing gas in a Geissler tube emits its own characteristic light, these tubes were used for the spectroscopic study of emission lines. For this purpose, Geissler's associate, the Bonn physicist Julius Plücker, increased the intensity of the light by narrowing the tubes considerably (see above and illustration at right). With these modified tubes, Plücker discovered the first three lines of hydrogen, before the more well-known efforts of Bunsen and Kirchhoff.

After Geissler's death in 1879, Franz Müller operated his shop. Müller's name appears on some of the Dartmouth tubes.

Geissler tubes can be very elaborate examples of the glassblower's art. The discharge makes the glass fluoresce; different glasses emit different colors. Glassblowers took advantage of this phenomenon to produce some spectacular designs. Because of their beauty, Geissler tubes are still produced. Modern neon signs and fluorescent lamps are practical derivatives of Geissler tubes.

Müller's 1904 catalogue offers a helium-filled tube, similar to ours, for 12.50 Reichmarks; the iodine-filled tube sold for RM 3.[116]

164

Electric egg (aurora jar)

APPS
London, England
1871
Height: 74 cm   2002.1.34739

In addition to providing emission spectra, Geissler tubes were used to model some natural phenomena. By the end of the eighteenth century, evidence had accumulated suggesting that the aurora borealis was produced by electricity interacting with the atmosphere. Apparatus for demonstrating this phenomenon became popular, and such "aurora jars" are Geissler tubes, usually egg shaped.

Unlike conventional Geissler tubes, however, electric eggs have a valve and can be evacuated to different pressures. Different gasses or even liquids can be introduced. Likewise the upper electrode can be raised or lowered to adjust the gap.

Using an electric egg, the Swiss physicist Auguste de la Rive first showed that magnetic fields alter the emission of light in a patially evacuated chamber. He used this phenomenon to explain the effect of terrestrial magnetism on the motions of the aurora borealis. Geissler later constructed a special tube to demonstrate the effect. This phenomenon would be explored in considerable detail with cathode ray tubes.

In February 1871, Professor Charles A. Young purchased this egg from Alfred Apps for £4.[117]

CROOKES TUBES

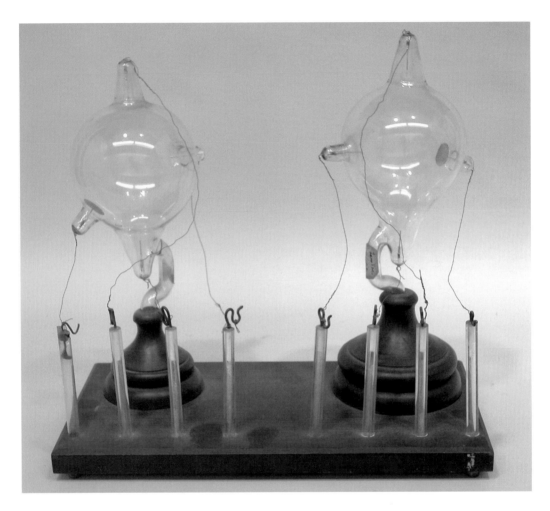

Unknown maker
Late nineteenth century
Height: 32 cm

2002.1.34713

When the vacuum within a Geissler-type tube is increased, the phenomena exhibited change dramatically. Such high-vacuum tubes are called "Crookes tubes." Whereas in the Geissler tube the electrical charge is carried on a plasma (the ionized residual gas in the tube), when the vacuum is increased a dark space, termed "Crookes dark space," appears in the plasma near the cathode. At even higher vacuum, the plasma disappears because there are too few gas molecules to form a conductive medium.

The British chemist William Crookes began exploring these phenomena in 1876. He designed apparatus that could evacuate tubes to different pressures, filled with different residual gases. At lower pressures, he found that the cathode emissions follow linear paths, are deflected by a magnet, carry negative electrical charge, and cast shadows of opaque objects. The German physicist Eugen Geissler also studied these emissions in 1876, naming them "cathode rays." Deeply interested in Victorian Spiritualism and its fashionable séances, Crookes speculated that the rays are corpuscular, consisting of what Michael Faraday had called the "fourth state of matter," a "radiant matter" that exists on the border between "regular" matter and force.

166

The wood engraving, reproduced below from Adolphe Ganot's *Elementary Treatise on Physics* (1890), shows a pair of Crookes tubes similar to the Dartmouth set.[118] The tube on the right is more highly evacuated than that on the left; the illustration shows the characteristic path of the rays. The arrangement shown here allows the investigator to apply different voltages to different electrodes in the tubes.

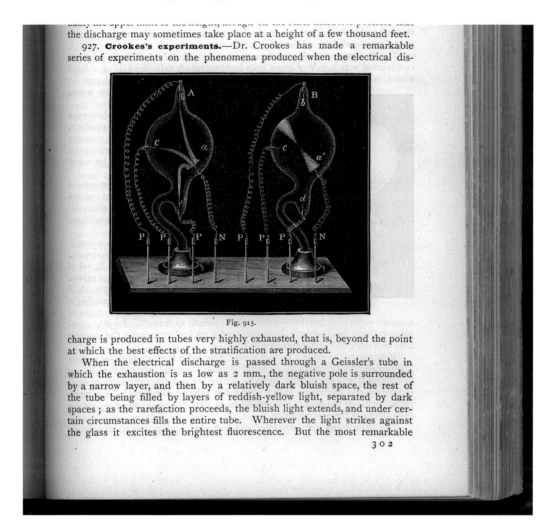

the discharge may sometimes take place at a height of a few thousand feet.

927. **Crookes's experiments.**—Dr. Crookes has made a remarkable series of experiments on the phenomena produced when the electrical dis-

Fig. 915.

charge is produced in tubes very highly exhausted, that is, beyond the point at which the best effects of the stratification are produced.

When the electrical discharge is passed through a Geissler's tube in which the exhaustion is as low as 2 mm., the negative pole is surrounded by a narrow layer, and then by a relatively dark bluish space, the rest of the tube being filled by layers of reddish-yellow light, separated by dark spaces ; as the rarefaction proceeds, the bluish light extends, and under certain circumstances fills the entire tube. Wherever the light strikes against the glass it excites the brightest fluorescence. But the most remarkable

3 0 2

The pair of Crookes tubes in the King Collection is mounted on turned pedestals of different sizes. The base as well as the glass insulators have a homemade appearance, suggesting that this instrument was assembled locally from separate components. Perhaps this instrument was constructed after the one illustrated in Ganot's textbook, a copy of which was located in the Physics Department during the 1890s.

Editions of Ganot's book from the 1880s-90s are filled with descriptions of the multitudinous phenomena of cathode rays. The discovery of X-rays (1895) and the electron (1897) depended upon the equipment of cathode ray physics—vacuum pumps, evacuated glass tubes, methods for sealing electrodes in glass, and high-voltage sources such as induction coils—that had been widely available for the previous two decades.

Helium X-ray tube

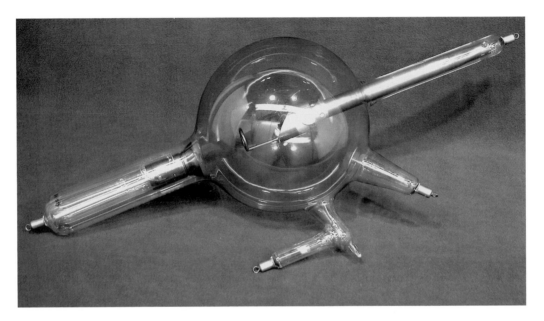

Victor, manufactured by Macalaster Wiggin Co.
Boston, Massachusetts
About 1917
Length: 55 cm
Exhibited in "Illuminating Instruments," Hood Museum of Art, Dartmouth College, 2004.   2002.1.34676

A Crookes tube was instrumental in the discovery of X-rays. In 1895 Wilhelm Röntgen, a physicist at the University of Würzburg, covered one of his Crookes tubes with black cardboard, but noted that a screen, some distance away, coated with a fluorescent salt, glowed faintly whenever he operated the covered tube. He systematically studied the phenomenon and discovered that the unknown radiation could pass through many substances, including human flesh, and referred to it as X radiation. Early X-ray tubes were, in essence, low-pressure, gas-filled Crookes tubes. In the tube above, a high potential (generally 30 to 50 kilovolts) between the cathode and anode (target) causes ionized helium to smash into the cathode, ejecting electrons which in turn strike the tungsten anode with energy sufficient to excite its atoms to emit X-rays.

The tubes were most often powered by induction coils or electrostatic machines such as Wimshurst generators. Apparatus, such as the Apps Ruhmkorff coil, which had, but a few years earlier, been sold simply as an induction coil was now being marketed as "Röntgen Ray Apparatus."[119]

X-rays rapidly found practical use in many areas, especially medicine. Medical instrument makers quickly added X-ray equipment to their offerings, as can be seen from the following 1901 catalogue pages of Allen & Hanburys, a leading London maker of medical and surgical equipment.

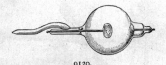

9120.

No.		£	s.	d.
9120.—**Focus Tubes,** with platinum anodes, mounted on aluminium, exhausted for use up to 6-inch spark, each		1	1	0
9121.—Ditto ditto ditto for 6- to 10-inch spark, each...		1	5	0
9122.—Ditto ditto with osmium-iridium anodes, mounted on aluminium, specially constructed for use with large coils, each		2	10	0
9123.—Ditto ditto ditto mounted between nickel and platinum, and with special arrangement for quickly lowering the vacuum when necessary		3	3	0

These tubes are specially constructed for use with the electrolytic brake, working with current direct from the main.

The tubes Nos. 9122 and 9123 are focussed to a great nicety, and give absolutely sharp definition on the photographic plate.

All of the above tubes are of English manufacture, and for high-class skiagraphy they are unequalled.

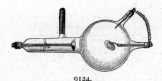

9124.

		£	s.	d.
9124.—**Focus Tubes,** with double anodes, specially suitable for screen work, for use with any coil up to 6-inch spark, each		1	1	0
9125.—Ditto ditto highly exhausted and specially suitable for use with large coils ...		1	5	0

These tubes give very brilliant results on the screen, and have a high degree of penetration.

Every tube supplied by Allen & Hanburys Ltd. is very carefully tested before being sent out, and is guaranteed to give the finest results.

From the nature of their construction, Focus Tubes are exceedingly fragile, and there is considerable risk of breakage in their transmission either through the post or by any other mode of conveyance.

Allen & Hanburys Ltd. use every precaution in packing, but they do not guarantee safe delivery, and can only supply them at the risk of purchasers.

Röntgen ray tubes and complete outfits were offered for sale quickly after the discovery of the rays, as seen in these pages from the 1901 Allen & Hanbury's catalogue.

Complete Röntgen Ray Outfits.

Suitable for Hospitals, Institutions or Consulting Rooms.

No.		£	s.	d.
9135.—6-in. Spark Induction Coil		15	10	0
12-Volt Accumulator		4	0	0
Fluorescent Screen, size 10 in. by 7 in.		2	5	0
Portable Fluoroscope		2	10	0
2 Vacuum Tubes		2	2	0
Universal Tube Holder		0	12	0
		26	19	0
9136.—10-in. Spark Induction Coil		27	10	0
12-Volt Accumulator		4	15	0
Volt Meter		3	10	0
Ampère Meter...		3	5	0
Fluorescent Screen. size 10 in. by 7 in.		2	5	0
Portable Fluoroscope		2	10	0
2 High Vacuum Tubes		2	10	0
Universal Tube Holder		0	12	0
		46	17	0
The above set with 10-in. coil, No. 9080, in place of coil No. 9084 extra ...		12	10	0
9137.—18-in. Spark Induction Coil		85	0	0
Two 12-Volt Accumulators		9	10	0
Volt Meter		3	10	0
Ampère Meter		3	5	0
Fluorescent Screen, size 15 in. by 11 in.		5	0	0
2 High Vacuum Tubes, with osmium-iridium anodes		5	0	0
Universal Tube Holder		0	12	0
		111	17	0

When any of the above sets are to be used with currents direct from the mains, special rheostats and other forms of breaks may be necessary. Special quotations upon application.

Complete Photographic Outfits for use with above sets supplied to order.

Estimates upon application.

53

This illustration, showing what is reputed to be the first medical X-ray in North America, has long been a central icon for the history of science and medicine at Dartmouth College.

In the early weeks of 1896, the scientific and popular press flashed news of Röntgen's discovery around the world. On January 26, the *New York Sun* carried a story entitled "The new photography" that prompted a Yale physicist the following day to use a Crookes tube from his collection to make the first recorded X-ray photograph in America. Howard H. Langill, a prominent local photographer in Hanover, New Hampshire, also read the *Sun's* story and urged a friend, Frank Austin, a recent Dartmouth graduate and assistant in the physics department, to test the dozen or so Crookes tubes at Dartmouth for the production of the new rays.

The investigators used Dartmouth's Apps induction coil (see entry, page 160) to generate high voltage from a battery of eight Grove cells (see entry, page 154). They found that a Puluj tube made by Emil Stöhrer of Leipzig, rather than a Crookes tube, produced the most abundant beam of rays, as accelerated electrons smashed into a thin sheet of mica placed obliquely within the tube. Puluj tubes had been used since the mid-1880s to demonstrate phosphorescence; no one suspected that they also had been producing copious amounts of X-rays.

On January 19, young Eddie McCarthy of Hanover had broken his arm while ice skating. His physician, Gilman D. Frost of the Dartmouth Medical School, was brother of the professor physics and astronomy, Edwin B. Frost. On February 3, Gilman took McCarthy to the physics department's apparatus room in Reed Hall, where they X-rayed the broken ulna with a twenty-minute exposure. The above illustration, probably a later recreation shot by Langill, shows McCarthy and his mother, with physicist Frost timing the exposure and physician Frost standing at the right.

Dated February 4, Edwin Frost's article showing several Röntgenograms (but only describing the broken arm without presenting its X-ray image) appeared in the February 14 issue of *Science*. The same issue includes articles by physicists from Columbia and the University of Pennsylvania, both dated February 8, that also discuss clinical X-rays. Well stocked with apparatus required for the study of cathode rays, many physics departments at American universities and colleges rapidly moved to study and exploit Röntgen's discovery.[120]

COOLIDGE TUBE

GENERAL ELECTRIC CO., manufactured by VICTOR
Schenectady, New York
1916-1920
Length: 55 cm   2002.1.34683

The Coolidge tube was a fundamental departure from the early Crookes tube-based X-ray tube. The Coolidge tube did not rely on a very large potential difference between cathode and anode. By increasing the vacuum within the tube and replacing the cathode with a flat spiral tungsten filament, William David Coolidge, a research physicist at GE, found in 1913 that he could heat the cathode to incandescence and "boil off" electrons. Because the tube had a high vacuum, the mean free path of the electrons was relatively long, increasing efficiency. And by removing the gaseous ions of earlier tubes, the frequency of the resulting X-rays became more precisely controlled (maintaining the gas pressure in earlier X-ray tubes had been notoriously difficult). The greater the voltage, the greater the acceleration of the electrons and the penetrating power of the emitted X-rays. Appropriately designed Coolidge tubes could be operated at voltage ranges from a few hundred to about a million volts.

This is an early Coolidge tube made by Victor for the General Electric Corporation (patented 1917). In 1920, GE would take complete control of Victor. The use of the Victor name and trademark, however, would continue until about 1930.

Contemporary medical authors widely praised the advantage of the "hot" Coolidge tube over earlier gas-filled tubes, citing longer tube life, constancy of tube performance, greater intensity or penetrating power, improved reproducibility of results deriving from enhanced control over the range of wavelengths and the intensity of the X-ray beam, self-rectification of the applied voltage, better focusing, and elimination of indirect rays. Such advantages soon made the Coolidge tube the main type used for medical diagnosis and treatment.[121]

The Coolidge tube quickly found its way into popular literature. In one 1939 novel, a character speaking in 1914 "hoots at the idea that we have to inquire of Germany for the latest tricks in applied science." He had just returned from a trip to Germany to examine their efforts at improving X-ray tubes, and came home to the Coolidge tube.[122]

172

In this detail, we see the tube within the thick-walled lead glass outer bowl that blocks stray X-rays. The cathode and the target are seen near the center. The target, a flat piece of tungsten at a 45^0 angle, directs X-rays toward the bottom of the tube where they pass through a hole. With an appropriate power supply, it was possible to accurately control the amperage supplied to the filament and the voltage between the anode and cathode. In actual use, the heat is carried away from the anode by the long copper tube.

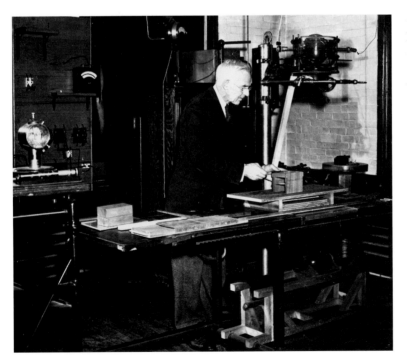

Dartmouth has a long tradition of work with X-rays. Both gas and hot cathode tubes were used by Professor Arthur Bond Meservey (Dartmouth, 1906) for teaching and research until 1950, and from 1913 to 1930 for diagnostic and therapeutic work at Mary Hitchcock Memorial Hospital, where he served as its first Röntgenologist. This illustration shows Meservey in the "X-ray room" on the third floor of Wilder Laboratory in 1947, preparing to use the Coolidge tube.

173

GOLD LEAF ELECTROSCOPE

W.G. PYE & CO.
Cambridge, England
About 1909
10 x 5 x 15 cm

2002.1.34920

Invented in 1904 by C.T.R. Wilson, the Cambridge physicist who had developed the cloud chamber and later would win the Nobel Prize in physics, this electroscope was designed to measure extremely slight changes in electric potential. A single gold leaf is suspended vertically from the upper electrode; a rectangular plate is attached to the other electrode and placed parallel to the back of the brass box. When tilted as shown, the geometry of the plate and leaf produces a characteristic voltage where the deflection of the leaf is maximally sensitive to differences in potential. In his original publication, Wilson reported that he could measure differences as small as 5 millivolts. Wilson and others used such electroscopes to study the production of ions by "atmospheric electricity" (cosmic rays) and radioactivity.[123]

Fig. 308. No. 8640.

ilted rectangular pattern (see Fig. 308), with ebonite the leaf, complete on stand with levelling screws, as Cavendish Laboratory, Cambridge ...   £2 . 0 . 0

Pye & Co. catalogue, 1911

174

Tangent galvanometer

J.N. Brown
Hanover, New Hampshire
About 1880
Height: 45 cm; Diameter of the coil: 35 cm

2002.1.34938

Early-nineteenth-century electroscopes used to measure the newly discovered "galvanic" current were called "galvanometers." After H.C. Ørsted's discovery of electromagnetism in 1820, the term soon became

175

reserved for current-measuring devices that employ coils and magnetic needles. Current flowing in the large coil of wire produces a magnetic field which deflects the needle. To use the tangent galvanometer, invented by the Parisian professor of physics Claude Pouillet around 1837, the coil is placed parallel to the local magnetic meridian. As current flows through the coil, a magnetic field is produced perpendicular to the coil, and the tangent of the needle's deflection is proportional to the strength of the current. This proportionality is relatively exact only if the length of the needle is less than about one-eighth of the coil's

Dovetail brass seam, John Brown,
about 1880

diameter. In our galvanometer this ratio is only about 1:3, which may have affected the instrument's accuracy. Because of their relatively few coils, tangent galvanometers have low resistance and thus are well suited to measure strong currents at low voltages. But with their coils so distant from the compass needle, tangent galvanometers are not very sensitive to small current changes.[124]

This galvanometer is one of the few instruments of the King Collection made locally by a craftsman not employed by Dartmouth College. A local gazetteer of 1886 described John N. Brown as a "machinist...[who has] carried on the business of making special machinery, models, and general jobbing [for] about twelve years." Brown's shop was located near Hanover's Main Street. In 1889, he sold the business and became "foreman of machine work" at the workshop of the New Hampshire College of Agriculture and Mechanic Arts, the state's land-grant college then located in Hanover.[125]

The brass cylinder is of particular interest because it was rolled and the joint soldered, the common way of producing cylinders from sheet stock. However, in this instrument, the maker used a dovetail joint (top, left) to join the seam near the bezel. The lower part of the cylinder has an ordinary butt joint. All the joints are soldered with what appears to be a hard solder, then burnished smooth.

Dentile copper seam, early-
twentieth-century Ottoman

The dovetail joint is unusual in metalworking. Increasing the strength of seams by means of interdigitating mating edges was a common practice among eighteenth- and nineteenth-century makers of brass musical instruments. Persian and Ottoman metalworkers also employed this technique. However, these metalworking traditions generally used dentile (box) rather than dovetail joints.[126]

Dentile copper
seam, early-
twentieth-century
Ottoman

176

SINE GALVANOMETER

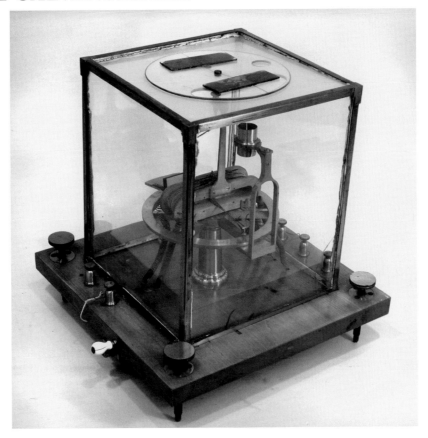

Breguet
Paris, France
1853
29 x 28 cm

2002.1.35198

Both the sine and tangent galvanometers were devised by the Parisian physicist Claude Pouillet and both are characterized by having a coil of wire around a centrally located magnetic compass. They are operated differently, however. In the sine galvanometer, the coil is first set parallel to the needle, hence parallel to the magnetic meridian. When current is applied, the coils are rotated until parallel to the needle, which has assumed a new azimuth and is now acted upon by both the earth's magnetic field and the magnetic field from the coil ("coupled"). The strength of the current is related to the sine of the angle, or rotation of the coil, needed to maintain this relationship. An advantage of the sine galvanometer is that the needle remains perpendicular to the field of the coil, i.e. in a non-uniform but symmetric magnetic field. In this geometry the needle can be large relative to the coil, and the sensitivity of the sine galvanometer is considerably greater than that of the tangent galvanometer. The above instrument has astatic needles (a pair of parallel magnetic needles mounted with their polarities reversed), which reduces the effects of the earth's magnetic field.

Louis Breguet was a pioneer in electrical telegraphy in France in the 1840s. This instrument, perhaps deriving from this delicate work, is one of the more distinctive in Dartmouth's large and varied collection of galvanometers. Utilizing a torsion-wire suspension, the instrument is in a glass housing to block air currents. It originally had a telescope to read the vernier, but the telescope is now missing. The brass feet are screws to facilitate accurate leveling. Apparently this design was quite successful. A Breguet catalogue from 1877 depicts an identical instrument, then available for £10.[127] Ira and Charles Young purchased this instrument in 1853.

FOUR-COIL ASTATIC MIRROR GALVANOMETER

Unsigned (possibly Heinrich Rubens)
Berlin, Germany
About 1895
Height: 54 cm
Exhibited in "Illuminating Instruments," Hood Museum of Art
Dartmouth College, 2004.   2002.1.35205

The design of this galvanometer, with tiny, astatic magnets suspended on a torsion fiber between two pairs of facing coils, was invented in 1859 by William Thomson specifically to detect extremely weak signals in submarine telegraph cables. For the next four decades, galvanometers made after Thomson's basic design were considered to be the most sensitive available and were widely produced in many forms.

By 1881, Siemens and Halske were selling a model somewhat similar to ours. Henri DuBois and Heinrich Rubens in 1892 further refined the instrument by making the coils easily removable and providing interchangeable magnet-mirror torsion elements of different masses. Thus, the instrument could be readily altered to minimize the effects of minute vibrations in the environment. According to their published table of sensitivities, DuBois and Rubens reported that they could measure currents of less than ten nanoamperes with 20-ohm coils such as are on ours.

The DuBois and Rubens design was manufactured by Keiser & Schmidt in Berlin. An illustration identical to that in their original article appears in Max Kohl's 1909 catalogue, but bearing the Kohl signature prominently.[128]

Given its similarity to the illustration in DuBois and Rubens' original publication and its German markings (terminals labeled *A[nfang]* and *E[nde]*), this galvanometer undoubtedly was made in Germany. In the absence of a signature, it is tempting to speculate that this galvanometer may not have been commercially made but rather originated in Rubens' laboratory where Ernest Nichols had worked from 1894 to 1896, before he came to Dartmouth.

In their 1901-03 experiments on the pressure of light, Nichols and Hull used this galvanometer. Unfortunately the magnet-mirror torsion element is now missing.[129]

EINTHOVEN STRING GALVANOMETER

CAMBRIDGE INSTR. CO.
Ossining, New York
About 1925
15 x 22 x 17 inches   2002.1.34932

The string galvanometer, first offered in 1905 by the Cambridge Scientific Instrument Company, initially had been designed by the Dutch physiologist Willem Einthoven to detect the extremely small variations in electric potential that accompany contractions of the human heart muscle. By 1908, CSI was offering complete electrocardiograph units and the equipment, like the X-ray machine, quickly spread with great fanfare to hospitals and clinics as a powerful, physics-based diagnostic tool. For his work in electrocardiology, Einthoven won the 1924 Nobel Prize in Physiology and Medicine.

Judged by a recent historian to be the "most sensitive electrical measuring instrument" devised by 1900, the string galvanometer replaced the moving coils, small magnets and mirrors of earlier galvanometers with a tiny quartz filament, silvered to make it conducting. To trace the heart's electrical behavior, Einthoven needed to detect tenths of millivolts, varying on time scales of milliseconds. CSI's first galvanometers had strings only 0.003 mm in diameter that moved only 0.05 mm when several millivolts were applied. A massive electromagnet produced a constant field. Minute currents in the fiber set up a magnetic field around the fiber, which caused it to deflect. A powerful optical system was required to magnify and record these minute deflections.

179

Einthoven's original galvanometer, employing an enormous water-cooled electromagnet, filled two rooms in Leiden's physiological institute and required five people to operate. Telegraph wires brought the signals from patients, located 1.5 km away in the university hospital, to the galvanometer. Einthoven did not patent his electrocardiograph. Rather, he licensed its production to CSI, who engaged the British electrical engineer, William Du Bois Duddell, to design a portable string galvanometer (ours weighs about 300 pounds). In the 1890s, Duddell invented a fast-acting oscillograph for the electrical industry to monitor rapidly varying voltages. CSI's string galvanometer thus brought together the technologies and demands of both electrophysiology and electrical engineering. Early purchasers of the CSI instrument included not only hospitals but also experimental physiologists, wireless telegraph companies, and other instrument makers.

Because of high import duties, in 1922 the Cambridge and Paul Instrument Company (as it was called between 1919 and 1924) was forced to look for partners in the United States. They signed an agreement with Charles Hindle, a mechanic in Ossining, New York, who had been making an instrument based on a version of Einthoven's design as modified by Horatio Williams of the Cornell University Medical College (separate toroidal electromagnetic units placed contiguously around the C-shaped core of the large magnet). In 1924 the company went public and was renamed the Cambridge Instrument Company. Our model derives from Hindle's design, which differs from those based on the British design.[130]

The small currents detected by this instrument are not derived from electricity, in its classical sense, but rather arise from differences in ionic concentration across semipermeable cell membranes that normally have high dielectric properties. The conduction of an impulse along a nerve is the result of localized changes in membrane permeability and is rather unlike electrical transmission along a conductor.

180

Ayrton's variable inductance

LEEDS AND NORTHRUP CO.
Philadelphia, Pennsylvania
About 1920
Height: 14.5 inches

2002.1.35105

   William Edward Ayrton, professor of physics and electrical engineering at the Central Technical College in South Kensington, and John Perry, his mechanic, developed this device in the 1890s. With coils wound on sections of concentric spherical surfaces, one can demonstrate and measure the self and mutual inductance of coils. Ayrton had studied under William Thomson (Lord Kelvin), and invented many portable electrical measuring instruments before his death in 1908.

   The two coils, "wound on laminated mahogany bobbins and assembled in a very substantial mahogany frame," according to the 1927 Leeds & Northrup catalogue, are connected in series. When aligned with their currents in opposite directions, the inductance of the combined coils is zero. At ninety degrees, their inductance is the sum of the self-inductance of the coils. When their currents flow in the same direction the total inductance is the sum of the self-inductance plus twice their mutual inductance. On the above instrument, angles between the coils can be read to 0.1 degrees by means of a vernier.

The 1927 catalogue lists this "inductometer" for $185, and notes that a separate "curve giving self inductance values at 1000 cycles" is included.[131] The latter is not extant in the King Collection.

Willis M. Rayton, a professor of physics at Dartmouth is shown (above) in a teaching laboratory, about 1950, with several students. In the 1940-50s, Rayton taught courses on electricity and magnetism, electronics and high frequencies, and radio instruments and measurements. He was well known as a radio enthusiast. The loop antenna, visible against the window at the far right and currently in the King Collection, was made by Rayton. Ayrton's variable inductor can be seen on the table.

One of these students (second from right), Pieter von Herrmann (Dartmouth, 1950), subsequently earned a Ph.D. in physics from Yale and spent his career designing nuclear power plants for General Electric.[132]

182

VARIABLE RESISTANCE COIL

Unsigned
First half twentieth century
Diameter: 30 cm

2002.1.35232

By connecting the coil in series and rotating the central pivot arm, which moves a sliding contact brush, the length of the spiral strip in the circuit is varied, as is its electrical resistance. The brush slides radially along the pivot arm to follow the changing radius of the coil.

The coil has four terminals, allowing connections to the center pivot, the inner- and outermost ends of the coil, and to its midpoint. A gear links the center pivot to a rotating, non-conducting shaft, seen protruding through the outer frame at the bottom of the photograph. Apparently this shaft once held a dial pointer to register the angular position of the center pivot.

The measured resistance of the entire strip is less than one ohm.

The coil and arm appear to be commercially made. However, they are attached rather crudely to a rough piece of dimensional timber. Between the wood and the bottom of the coil is another smaller piece of wood, on which a separate flat 6-coil helix of bare copper wire was placed. It appears as if the basic commercial instrument has been locally modified to make a transformer.

WHEATSTONE BRIDGE

Otto Wolff
Berlin, Germany
1899
14 x 23 x 45 cm
Exhibited in "Illuminating Instruments," Hood Museum of Art, Dartmouth College, 2004.   2002.1.35233

   The Wheatstone bridge is an electrical circuit that provides a sensitive means to measure an unknown resistance. First described in 1833 by Samuel Christie and more completely in 1843 by Charles Wheatstone, the bridge consists of three known resistors, an unknown resistor, a battery and a galvanometer. If the known resistances are varied until no current flows through the galvanometer (a null reading), the unknown resistance can be determined.

   The bridge shown above has terminals for connecting an external battery, galvanometer and unknown resistance. The bridge circuitry and resistors are inside the wooden case. By removing or inserting plastic-handled copper plugs, a known resistance of up to 50,000 ohms can be varied in increments of 0.1 ohms. The resistors are made of manganin, a copper-manganese-nickel alloy then commonly employed for such purposes.

   A brass plate at the left end of the case is stamped with the German eagle and *PTR.II 130**. Presumably Wolff sent the bridge to the Physikalisch-Technische Reichanstalt, Section II, for calibration. Located in Berlin, the PTR was the German equivalent of the United States National Bureau of Standards. Dartmouth physicists Ernest F. Nichols and Gordon Hull used this bridge between 1900 and 1903 in their experiments to measure the pressure of light. Nichols had spent two years (1894-96) working at the physics institutes of the University and the Technical University in Berlin and had developed extensive contacts with the German scientific community. Nichols purchased this bridge in January 1900 for RM463 ($111).[133]

184

STUDENT VOLT-AMMETER

L.E. KNOTT APPARATUS CO.
Boston, Massachusetts
About 1900
Height: 15.5 cm; Base: 10 x 17 cm   2002.1.35057

In 1884, the British instrument makers Ayrton and Perry announced the first "direct-reading" volt and ammeters. Using a permanent magnet with an adjustable charcoal-iron core, the meters required an initial calibration but then were easily portable and, according to Ayrton and Perry, offered a precision of one percent for measurements of up to ten volts or amps. Far less sensitive than laboratory galvanometers, the direct-reading instruments initially were marketed to electrical engineers for use in industrial settings.[134]

L.E. Knott's volt-ammeter, however, was intended for the student laboratory. The cast aluminum housing displays a pattern of stylized vines that reflects influences from Art Nouveau. This international movement, emerging in the early 1890s in France and Belgium and quickly spreading across Europe and the Americas, shaped decorative styles in many media, ranging from furniture, glass, graphic art and book design to architecture, metalwork and jewelry. Art historian Nikolaus Pevsner characterized the Art Nouveau style as "the sinuosity of vegetation-inspired floral forms." Its leitmotif was the "long, sensitive curve, reminiscent of the lily's stem."[135]

L.E. Knott did not long manufacture such ornamented instruments. Their 1916 catalogue shows instruments with far less decoration; the 1921 catalogue portrays only severe black boxes with unadorned black enamel finishes. An aesthetic of the Machine Age had replaced Art Nouveau.[136]

STUDENT GALVANOMETER AND LAMP
REFLECTING GALVANOMETER

LEEDS & NORTHRUP CO.
Philadelphia, Pennsylvania
About 1940

Stand and reading scale (above)
Height: 39 cm   2002.1.34912

Lamp for reading scale (below)
Length: 17 cm   2002.1.34942

Type R reflecting galvanometer (right)
Height: 23 cm   2002.1.34935

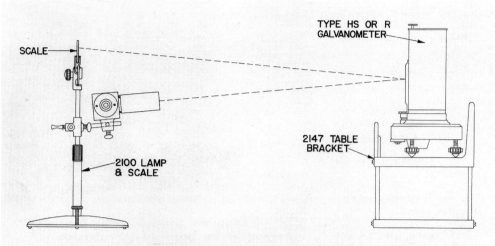

Fig. 2 - Arrangement of Galvanometer and Reading Device
for Horizontal Projection Method

The L&N plant above houses research, engineering and manufacturing departments as well as central offices for sales and service
at 4991 Stenton Avenue, Philadelphia

Founded in 1903, Leeds and Northrup were prominent makers of electrical measuring devices as well as photometric and ion-measuring equipment, such as pH meters. Their 1939 catalogue (next entry, page 188) lists twelve regional offices in the United States.

This student reflecting galvanometer (previous page, right) could be configured in several ways, one shown above. The galvanometer is fitted with a mirror attached to a moving coil. The movement of the mirror is amplified by the length of the reflected light beam. Expanding a scale by using a reflected light beam is a common design element of measuring instruments and is not restricted to electrical apparatus, but can be found on many other instruments, such as balances.

The above illustrations are from the instruction manual accompanying the student galvanometer.

187

STUDENT POTENTIOMETER

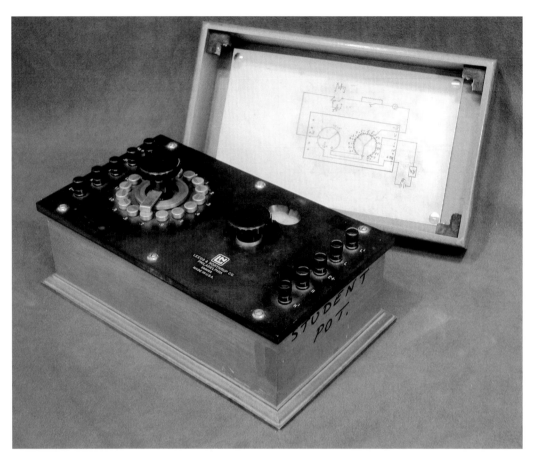

LEEDS & NORTHRUP CO.
Philadelphia, Pennsylvania
About 1926
16 x 20 x 36 cm

2002.1.35053

RTHRUP COMPANY

7651 STUDENTS' POTENTIOMETER

A simplified, moderate-precision potentiometer for educational and general laboratory use. Two ranges, 0 to 1.6 volts and 0 to 0.016 volt. Measuring circuit comprises 15 resistors of 10 ohms each with dial switch having exposed brush and studs, and a continuously adjustable slidewire of about 10 ohms. Slidewire scale divisions 0.001 volt on high range; 0.00001 volt on low range. Limit of error: dial resistors and total slidewire resistance alike within ±0.04 per cent; slidewire uniform within ±0.5 division. Grained bakelite top plate. Polished mahogany case with cover.....................................$70.00

Note: Separate connections to slidewire and end coils provided, so that slidewire of 7651 can be used in an a-c bridge (see page 32).

Leeds & Northrup catalogue, 1939

A potentiometer measures the potential of an unknown voltage source by balancing it against a known voltage. As designed by J.A. Fleming in 1885, the circuit employs two variable resistors that are adjusted until no current flows through a galvanometer. The potentiometer above contains only two variable resistors. As noted in the 1926 Leeds & Northrup Bulletin 765: "The circuits within the potentiometer have been made simple enough to be clear to students, and all accessories such as keys, switches, galvanometer, standard cell, etc. must be connected in circuit by the student. This procedure makes the student familiar with the potentiometer circuit."[137]

188

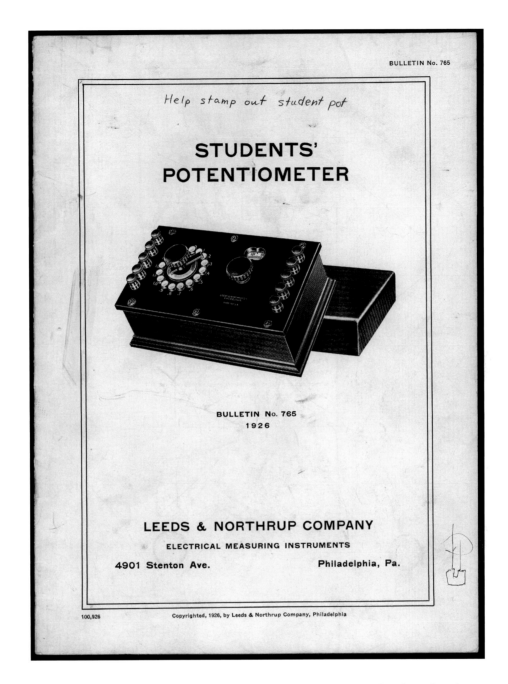

Help stamp out student pot

STUDENTS'
POTENTIOMETER

BULLETIN No. 765
1926

LEEDS & NORTHRUP COMPANY

ELECTRICAL MEASURING INSTRUMENTS

4901 Stenton Ave.   **Philadelphia, Pa.**

The bulletin describes how the potentiometer may be used to measure not only voltages but also currents and resistances by using the devices of the Wheatstone bridge. It also provides illustrations of the accessories required for measuring temperature with thermocouples, hydrogen-ion concentrations, and electrolytic conductivities. The potentiometer alone is listed for $80, a handsome sum for student laboratory apparatus in the 1920s. Potentiometers are also called voltmeters.

The instrument's design was successful. An apparently identical device is described in their 1939 catalogue and listed for $70, down from the 1926 price. Indeed, judging from the (presumably student) manuscript addition to the 1926 bulletin, this potentiometer must have been used at Dartmouth well into the 1960s.

Van de Graaff machine

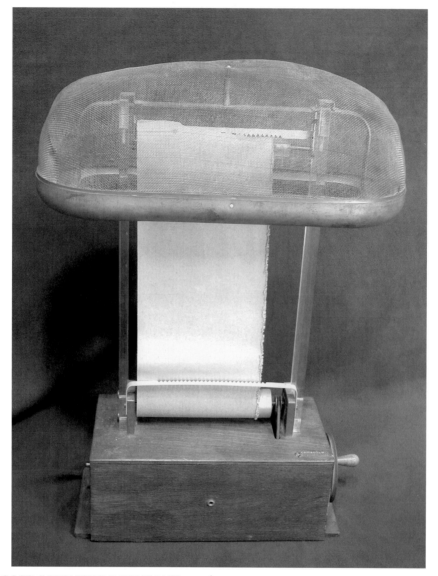

CAMBOSCO SCIENTIFIC COMPANY
Boston, Massachusetts
About 1950
Height: 57 cm

2002.1.34900

    The Cambosco catalogue calls this small Van de Graaff generator "Firsturn" and lauds its simplicity and reliability. "At the first turn of the crank, this modern Electrostatic Generator emits a crashing spark, whether the day is dry or humid! For that reason alone, it will be welcomed by every physics teacher who has ever apologized for the temperamental performance of an old-fashioned 'static machine'."

    With no Leyden jars, brushes, plates or collectors, this machine employs a hand-cranked belt to carry charge from the lower hollow metal terminal to the upper bronze mesh dome. According to the catalogue, which lists the device for $88.50, potentials of up to 300,000 volts can be attained without "the slightest hazard to operator or observer."

190

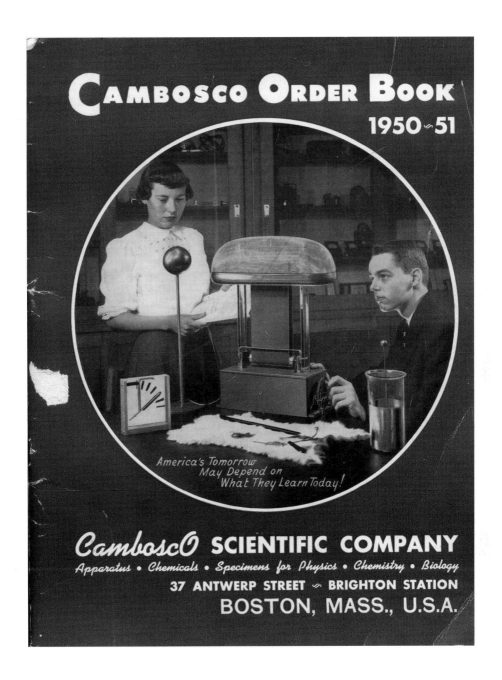

The emerging Cold War has influenced the rhetoric of this catalogue. Its cover features the generator and warns that "America's tomorrow" may depend on science education. The catalogue also noted that the Firsturn "enables you not only to perform the classical experiments in static electricity, but also to demonstrate the modern method of building up tremendously high voltages for atomic fission, nuclear research, and radiation therapy."[138]

Developed in the early 1930s at Princeton by R.J. Van de Graaff, these generators by the mid-thirties were producing potentials exceeding a million volts. Such high-voltage sources powered the particle accelerators for the rapidly emerging field of nuclear physics.

The Firsturn may have been the first Van de Graaff machine designed for general student use.

191

Edison lamps

EDISON'S PATENTS
(Left)
About 1883
Length: 4.5 inches   2002.1.34642

(Right)
About 1883
Length: 5 inches   2002.1.34640

(Center)
About 1888-94
Length: 5.25 inches   2002.1.34641

All exhibited in "Illuminating Instruments," Hood Museum of Art, Dartmouth College, 2004.

J.W. Swan and C.H. Stearn in England and Thomas A. Edison in the United States competed in developing a practical electric light source, culminating in Edison's 1879 patent of the high-resistance carbon-filament lamp. The commercial production of incandescent lamps required the convergence of several technologies, including the vacuum pump to evacuate the glass lamp. The small tips on the tops of the lamps result from fusing the evacuation port when the lamp is evacuated. Initially mercury gravity pumps were used, essentially creating a Toricelli space. These were supplanted by rotary mechanical pumps. The manufacture of a consumer commodity such as lamps transformed scientific apparatus into machines for commercial purposes.

The middle lamp (2002.1.34641) is 100 volt, 20 candle-power (c.p.). The extra threads and collar at the base date it as being somewhat more recent than the others. The plaster insulator is marked in pencil "Hanover, 100." The other two lamps are hand assembled and marked in ink *52 volts, 8 c.p.*. The filament connectors and stem press date them to 1883.[139]

The lamps shown here are some of the earliest to come from Edison's Lamp Works at Menlo Park, New Jersey. Charles F. Emerson, Appleton Professor of Natural Philosophy at Dartmouth, demonstrated them, using batteries as power sources, at the Dartmouth Scientific Association meeting on March 28, 1883. During the 1870s-80s, Emerson traveled widely through New England demonstrating Dartmouth electrical apparatus (the telephone, the light and the motor) and comparing their costs.[140]

STROBOTAC

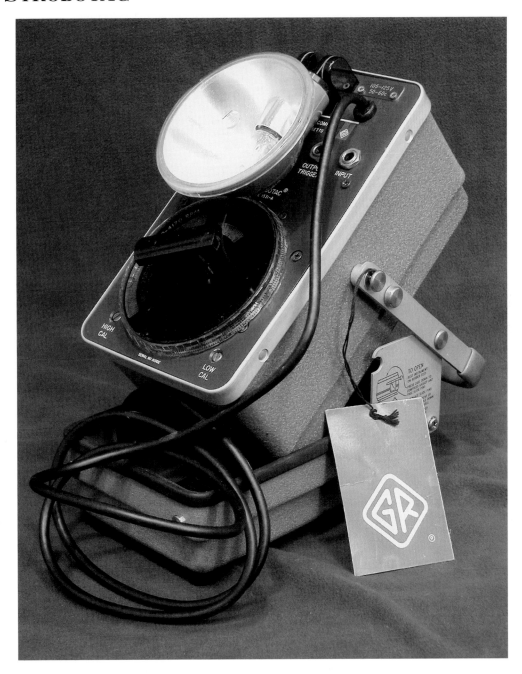

GENERAL RADIO COMPANY
Concord, Massachusetts
About 1963
Height: 17 cm

2002.1.34796

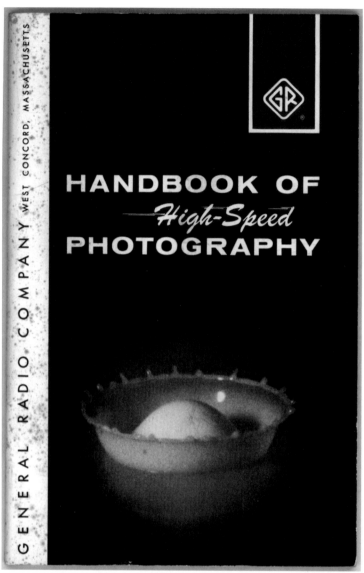

Strobotac instruction manual, 1963

Purchased by physics professor Francis Sears for use in classroom demonstrations, the Strobotac can track trajectories of rapidly moving objects (drum heads, falling drops of water, swings of golf clubs) by stroboscopic flashes. The popular physics textbooks by Sears and Zemansky include many "multiflash" pictures.

The technology of high-intensity light flashes had been developed to the point of practical use by Harold "Doc" Edgerton, Sears' former colleague at MIT. Since 1935, General Radio had been manufacturing Edgerton's first Strobotac. The new model, Type 1531-A, introduced by General Radio in 1960, featured a novel xenon lamp, developed by the firm Edgerton, Germeshausen & Grier, that produced flashes ten to twenty times shorter than the earlier model. The 1531-A generates flashes 1-3 microseconds in duration, at a rate continuously variable from 110 to 25,000 flashes per minute.

A comprehensive manual accompanies the instrument, introducing users to both technical and aesthetic aspects of high-speed photography. Company literature launching the Type 1531-A emphasized industrial applications of the strobe. Among other things, it could be used to investigate rotational and vibratory motions of internal combustion engines, fans and blowers, tape recorder heads, speaker coils, fuel-spray

194

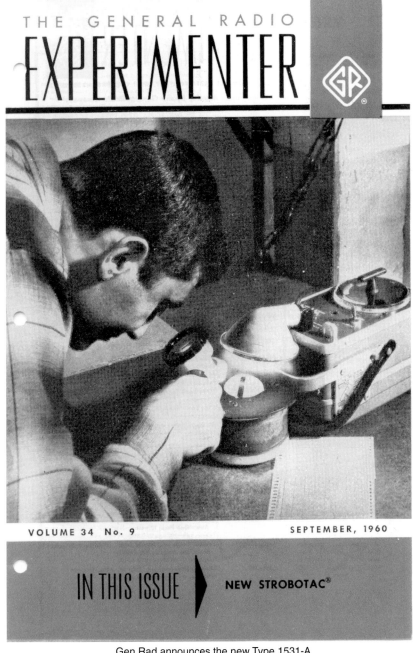

THE GENERAL RADIO

EXPERIMENTER

GR

VOLUME 34 No. 9   SEPTEMBER, 1960

IN THIS ISSUE ▶ NEW STROBOTAC®

Gen Rad announces the new Type 1531-A

nozzles, and turbine blades. It could check color registration of rotogravure printing presses. Even medical applications were mentioned.

In 1960 General Radio sold the new Strobotac for $260. The Strobotac line was sold in 1991 to QuadTech and later acquired by IET Labs, who currently offer a very similar Type 1531-AB strobe, at a list price of $4500.

The use of short-interval light flashes to "stop" motion dates at least to the 1850s. A disc with segments colored with prismatic hues appears dull grey when spinning rapidly. In a darkened room, the discharge of a Leyden jar will "stop" the motion and make visible the individual colored segments.[141]

CATHODE-RAY OSCILLOSYNCHROSCOPE

BROWNING LABORATORIES INC.
Winchester, Massachusetts
About 1940s
38 x 41 x 26 cm
Serial number: 523

By the 1870s, Geissler tubes had spread widely throughout the physics community. The rays passing between the electrodes in these tubes had become known as cathode rays and it had been discovered that magnetic fields bend these rays (see Crookes tube entry, page 166). Throughout the 1890s, however, attention remained focused on the characteristics of cathode rays themselves, rather than on their use for other purposes. This would change in 1897 when the Strasbourg physicist, Ferdinand Braun, realized that by placing electromagnets around a cathode ray tube, he could employ the deviation of the cathode ray to trace rapid variations in a current going through the electromagnet. Thus was born the first electronic recording system, representing an alternative to the electromechanical systems of rotating drums or coils that had spread widely through nineteenth-century experimental physiology, physics, and psychology.[142]

Braun's device, however, required considerable improvements before it became a practical instrument. In 1899, his student, Jonathan Zenneck, introduced a horizontal time scan controlled by a mechanically driven rotary potentiometer. Not until the 1930s did commercially manufactured, fully electronic oscilloscopes become available, employing vacuum tubes for the power supplies and other circuitry. Yet their horizontal sweeps could deliver only regularly spaced, "linear ramp" patterns, which meant that only repetitive signals of regular shapes could be displayed.[143]

During World War II, engineers at Los Alamos and MIT's Radiation Laboratory developed "synchroscopes" to display pulses from radar equipment and particle detectors. The timing circuits in these instruments could produce single horizontal sweeps, triggered by a selected level of the incoming signal, a simultaneous pulse from another counter or from other parts of a circuit (see the dials for *INT[ERNAL] SWEEP SELECTOR, SYNC GAIN* and *SYNC COLLECTOR* [not visible]). Synchroscopes can thus look at non-repetitive signals that are random in time and in shape. Contemporary observers lauded the flexibility of the new technology. A 1948 book from the Radiation Laboratory announced that "the presence of these features on the synchroscope render the oscilloscope obsolete except for very limited applications."[144]

However, if the new technology triumphed, the name did not. This oscillosynchroscope was used at Dartmouth in the 1950s for laboratory courses on Electricity and Magnetism. New scopes with identical features, purchased after 1960, were simply called oscilloscopes.

Around 1950, these students were experimenting with vacuum tubes, at the dawn of the transistor age. Note the large-scale diagram of a cathode-ray tube and the oscilloscope being set by one of the students.

Helium-neon laser

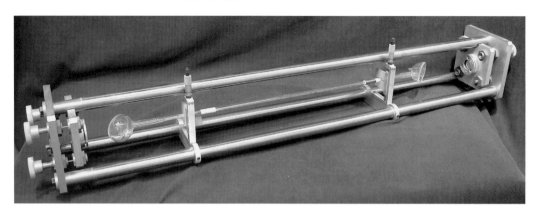

MASER OPTICS INC.
Boston, Massachusetts
1963
17 x 15 x 115 cm
Exhibited in "Illuminating Instruments," Hood Museum of Art, Dartmouth College, 2004.   2002.1.35327

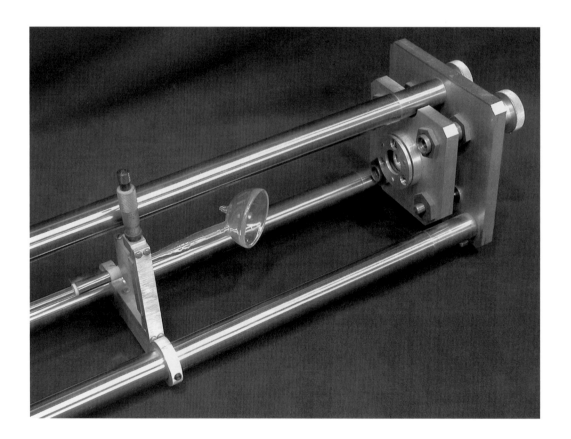

Devices for producing coherent radiation by stimulated emission were developed in the 1950s. Early masers, which yield radiation in the microwave range, used ammonia or ruby as the active medium. The first operating laser, which extended the maser principle into the infrared and optical range and also used ruby, was announced in 1960 by T.H. Maiman at Hughes Aircraft Company. Several months later, A. Javan, W.R. Bennett, Jr. and D.R. Herriott at Bell Telephone Laboratories obtained laser emission in a mixture of helium and neon, pumped with radiofrequency (RF) electrical discharge. MASER is an acronym for microwave amplification by stimulated emission radiation. LASER is the same with "light" replacing "microwave."

Relatively easy to construct, helium-neon lasers soon became commercially available. Allen King, who in the early 1960s was teaching an advanced course on optics, wanted to purchase such a laser for student use but only when the price fell below $1000. By early 1963, Maser Optics advertised a model for $1600 and a "cheaper" version for only $995. King immediately ordered the less-expensive model; but when the company proved unable to deliver it, they agreed to provide King with the more expensive model at the lower price.

This laser is powered by an external RF generator, with its leads attached to electrodes wrapped around the outer surface of the quartz resonating tube that is filled with helium and neon. The windows at the end of the tube are set at Brewster's angle, which partially polarizes the light passing through the windows to the external mirrors. The RF pumping excites the helium which in turn excites the neon. As the neon atoms return to their ground state, they emit photons which, in bounding back and forth between the external mirrors, prompt the emission of additional photons of the same wavelength and in phase (coherent). One of the external mirrors is only partially reflective, permitting the exit of a continuous beam of monochromatic (red, 632.8 nm), polarized, coherent light.

According to King, students employed this device to observe not only optical effects such as diffraction and interference, but also the mode structure, speckling and other characteristics of the coherent light beam.[145]

Maser Optics also provided a filmstrip and LP record with this laser. A spokesman explains the theory and operation of the ruby and helium-neon lasers, describes experiments students can perform, and outlines potential applications in communications, radar, materials fabrication, and surgery. With "faster growth than the transistor," gushes the spokesman, the laser will "revolutionize many technical fields." Not an instruction manual, the LP and the filmstrip provide a simplified description in multimedia. No other instrument in the King Collection is accompanied by such materials.

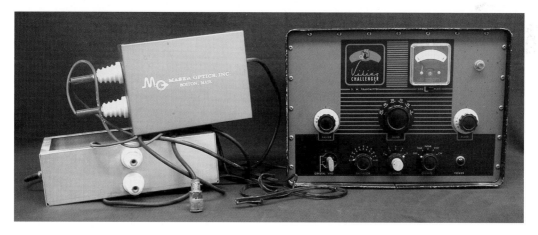

Radio frequency generator for the helium-neon laser
E.F. JOHNSON COMPANY
Naseca, Minnesota
About 1963
33 x 23 x 25.5 cm

2002.1.35328

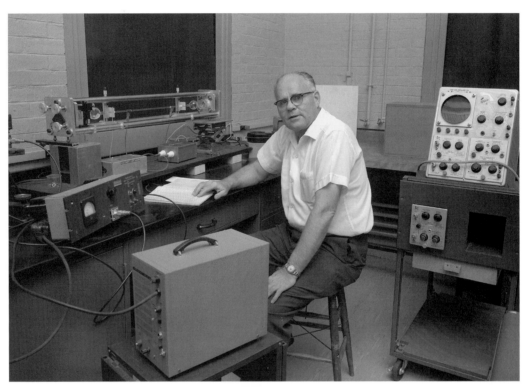

Allen King in Wilder Laboratory, 1966

Entering the storeroom, 2005

Chapter 1

The Cultural Attic

A visitor to an unmarked storage room in the basement of Dartmouth's physics building will find aisles lined with dozens of cabinets and open shelves, all overflowing with old scientific equipment. A chest-high glass disc mounted on a fancy wooden frame rests beside the door. Electrical equipment with art deco panels sits in a corner. A long, cast-iron bench signed "Franz Schmidt & Haensch, Berlin" blocks the entrance to a number of cabinets. A cabinet overflowing with early radio equipment emits the smell of aging electrical insulation. The room contains thousands of individual objects. Some, handsomely constructed of brass and glass, gleam richly through their age-toned varnish, displaying a proud patina. Others appear quite crude, made of wood with nails or screws and rough-sawn edges. Still others seem to be orphaned parts, isolated fragments of once larger, more complex objects. And one can find wooden cases, designed to hold and protect delicate objects but now empty. What unites these disparate objects? How did they find their way into the storage room? Is this a collection assembled for specific purposes? A museum?

The room also contains bookshelves filled with trade catalogues from instrument dealers and science textbooks. Several filing cabinets overflow with documents—invoices, instruction manuals, correspondence between professors and instrument manufacturers, historical notes on the local use of earlier instruments written by later Dartmouth professors, photographs—related to the instruments, their acquisition and use. Student laboratory notebooks, faculty lecture notes and earlier inventories of the instruments can be found in the College Archives. These documentary materials, like the instruments, appear to have been randomly aggregated.

From this room it would be difficult to assemble a coherent set of objects to demonstrate, say, the development of the microscope or the history of the radio. Nowhere in the cabinets could we find enough representative artifacts to display the chronological evolution of given instruments or technologies. Indeed, it would be difficult to identify a conceptual center of gravity for the items in our storage room. That is, the "old scientific apparatus" found at universities and colleges differ fundamentally from scientific instrument collections assembled by individual collectors or museum curators. To collect requires, in the words of a philosopher of hoarding, a "creative competence, a world-creating and world-ordering competence."[146] Collecting, either by individuals or institutions, needs criteria, strategies and interests. Collections have centers of gravity. Our storeroom does not.

Storerooms like ours, filled with old scientific instruments once used at universities or colleges, are cultural attics. They are places to which the everyday objects of earlier academic activity have been relegated, instead of being thrown away. Most of the objects in our room were initially acquired to study, measure or experiment; they were enlisted to teach about or to investigate the natural world. Over time, as the artifacts became less used, successive generations of

203

professors, laboratory assistants or janitors chose to push some of them further back on their shelves. Sometimes notes would be added, describing how the apparatus had been used locally. What we now call the King Collection of Scientific Instruments grew by undirected aggregation over the past two centuries, not by any deliberate process of collecting.

The first move to stabilize Dartmouth's cultural attic of scientific instruments and related documents came in 1960, when sixty old instruments were formally transferred

from the Department of Physics to the College Museum. The museum's curator of anthropology arranged for a student intern to inventory the acquisition, thereby creating the first museological record of what would become the King Collection. Sending the list to Allen King, then senior member of the physics department, the curator pined that "some day I hope we can find a student who would be interested in curating our old scientific instruments." In the same year, forty-seven pieces of surveying apparatus also came into the museum from an attic at Dartmouth's Thayer School of Engineering.[147]

Professor King, who had joined Dartmouth's faculty in 1942, would take up the curator's challenge. Already in the late 1950s, he had become curious about the "old apparatus" lying around his department and had begun to worry about its future in a building where space was getting scarce. When in 1964 the college celebrated the four-hundredth anniversary of Shakespeare's birth, King organized an exhibit "The Scientific Revolution, 1550-1850." Featuring dozens of instruments and books, this show represented the first public display, outside the physics department, of items from the cultural attic.[148]

By 1966, King systematically began to search the campus for other historical instruments. College administrators authorized two basement rooms in dormitories for storage. And memos began circulating about the possibility of assembling a large instruments exhibit for the college's upcoming 1969 celebration of its bicentennial. Deans found funds for a "Summer Project on Old Apparatus," supporting King and a student assistant as they cleaned and repaired artifacts, launched research on various types of instruments and instrument makers, and opened conversations with curators at national instrument collections and interested historians of science.[149]

The bicentennial exhibit, "Historic Philosophical Apparatus," curated by King and his son Ron, drew wide attention to Dartmouth's historic instruments. Supported by a grant of $10,000 from the National Science Foundation and sited in a central art gallery, the show attracted nearly ten thousand visitors. A press release described the genesis of project:

> For the last few years Professor King diligently retrieved glass, brass, and unidentified objects headed for the scrap heap, searched college property for the unnoticed but intact, and borrowed the revered items from academic departments in an effort to piece together as much as possible of the physical evidence of Dartmouth's science scholarship. His results, shown in this exhibit, are as much a chronicle of New England resourcefulness and of interest to the casual historian as they are to the scientist.[150]

By the time he retired from active teaching in 1975, King was ordering the attic. Named Adjunct Curator of Historical Scientific Apparatus at the College Museum, King over the next years opened a card catalogue of the objects and sorted them into functional categories. He secured the basement storage room (back in the physics department), made items accessible to visiting scholars, and arranged for loans to outside exhibits. He mounted small exhibits in the physics building and published a number of short articles on various instruments. And he drafted proposals to publish illustrated catalogues, offer a course on historic instruments, and create a budget and paid curatorial position for the collection. Although none of these latter ideas would take root, King nonetheless succeeded in transforming Dartmouth's long aggregated, retired apparatus from their endangered status in limbo into a semi-formalized collection. At Professor King's death in 2003, the materials he organized were named the Allen King Collection of Scientific Instruments.

Ironically, such a recent formalization of the collection helped to prevent its contamination by artifacts not related to Dartmouth. Over most of its history, the profile of Dartmouth's "old apparatus" remained so low that neither collectors nor alumni were tempted to contribute outside historic instruments to the college. Hence, the overwhelming majority of materials in the King Collection have a Dartmouth provenance; nearly all were used for science at the college.

Hundreds of similarly aggregated attics of historic scientific instruments can be found at universities across the world. From an abandoned nineteenth-century chemical laboratory in the middle of historic Athens (the Palaion Chimion) to a demonstration prep room at Columbia University in New York City, a vast landscape of the material culture of past science remains often hidden from view.[151] Visiting these spaces can be like finding unexplored archaeological sites; they preserve, *in situ*, the material culture of past science. At a few of the older institutions—Oxford, Cambridge, Padua or Harvard—the instruments, many collected from outside the university, have for decades resided in well-organized museums, replete with curators, budgets and catalogues. But at most universities and colleges, the retired instruments remain in their attic spaces, endangered and uncared for, not yet having found their Professor Kings.

What gets saved and why?

When a scientific instrument becomes obsolete–when it no longer can give sufficiently accurate data, or when all the data it can give are already known, or when a newer or different instrument can give data more cheaply–it only rarely goes directly into the tip. Usually it sits, retired, in some corner of the laboratory. Then perhaps it moves to storage and only gets thrown out when space becomes too dear or the department moves into a new building. But why save obsolete apparatus at all? A systematic history of instrument saving, hoarding, and aggregation remains to be written; we can only offer some impressions, derived from our studies of old instruments in university settings.[152] Regardless of its explanation, however, it is clear that without a process of aggregation, no university instrument collection could emerge.

Frugality, we speculate, may prompt professors to retain obsolete instruments near their original locations. A retired piece might be temporarily returned to service should the new model malfunction or break. Or retired apparatus might be kept to supply spare parts for later experiments, a pattern King himself observed at Dartmouth in the 1950s as new professors cannibalized old instruments. Just as machinists keep scrap bins in their workshops, so too do scientists keep old apparatus, because "you never know when you might need it."

Other instruments may be retained for personal reasons quite removed from the original functions of the objects. People become emotionally attached to things with which they have lived for extended periods. One of us (FJM) has fond memories of the many hours he spent using a Beckman Model DU spectrophotometer; he is looking for one to restore despite its lack of aesthetic appeal. Scientists may become sentimental about apparatus that they personally built or that helped them do a particularly significant piece of work. The tactile or aesthetic qualities of an instrument might make it difficult to discard. Many scientists and engineers, now retired, can lovingly show you the slide rules they used during their careers in the 1940-60s. Scientists develop a fondness for old instruments they have used, much as one develops fondness for an old dress, book, or shotgun.

Fame may give value to objects, encouraging people save them. George Washington's spectacles or Ginger Rogers' dancing shoes have found their way into national museums. Scientific instruments once used by Nobel Prize winners can attain a similar status. Fame tends to expedite an instrument's passage from back storage into collection, often not local. In such cases, the old instruments are kept for their totemic value, just as we preserve classic imprints in our libraries, classic architectural monuments on the National Register of Historic Places, or James Watt's steam engine in the Science Museum of London.

Even common objects of science are often retained. They need not be especially famous; they need not be beautiful, expensive or elegantly designed; they need not have been used by the professor doing the aggregating. But they do need a local provenance. They need to evoke stories of how science had been practiced earlier at the university where they continue to reside. Like the historical artifacts of everyday life preserved at local historical societies, the artifacts of everyday science aggregated by the Professor Kings enable universities to live in continuity with their past.

We often suggest to our undergraduates that they examine Dartmouth's copy of William Shakespeare's first folio, hold it in their hands and turn the leaves. Such tactile experiences form a cultural continuum that connects us to our human forebears in a way that no replica or digital image can. This linkage is broken when artifacts are discarded.

Chapter 2

An American College Gets Instruments

The history of scientific instruments in the academy follows different rhythms from those found in most college histories, organized around successive generations of presidents, trustees and administrators, anniversary celebrations, or academic departments. Instruments take their investigators to stories of entrepreneurial professors and students, self-images of educational institutions, commercial instrument makers and local craftsmen, practices of teaching and research, buildings, budgets, and patrons. Instruments involve patterns of acquisition, use, reconfiguration, replacement and loss. And scientific instruments, at Dartmouth College, will reveal themes that echo those occurring at many other American colleges and universities. In this sense, a history of Dartmouth's scientific instruments is a history of science in America.

Early acquisitions

Before the Revolutionary War, nine colleges had been established in the American colonies. After three early, scattered foundings—Harvard in 1636, Yale in 1701 and William and Mary in 1710—the pace quickened by the 1740s, with the appearance of the College of Philadelphia (later University of Pennsylvania), the College of New Jersey (Princeton), King's College (Columbia), the College of Rhode Island (Brown), and Queen's College (Rutgers). Dartmouth College, founded in 1769, was the last of these colonial institutions to be created.

Despite some variation in size and denominational control, by the 1770s most of these early colleges were quite similar. They sought to educate a male citizenry that would be both Christian and enlightened. Alongside a classical curriculum grounded in Greek, Latin, Hebrew, philosophy, divinity and ethics, the early colleges offered slight amounts of mathematics and geometry, astronomy and natural philosophy, and (occasionally) chemistry. Despite chronically precarious finances, the colleges generally sought to provide some pedagogical infrastructure to augment classroom instruction and recitation. Libraries, museums of "curiosities" (minerals, natural history specimens, archaeological artifacts), and collections of "philosophical apparatus" became sources of competitive pride as the early colleges struggled to attract students, benefactors and local boosters.

Dartmouth's founding president, a Congregational minister named Eleazar Wheelock, literally carved the college out of the wilderness of New Hampshire. Arriving in Hanover in the summer of 1770 with thirty students, Wheelock faced a tract of land covered with unbroken pine forest. Much of the first year they spent clearing land, digging wells, preparing gardens, building a sawmill, and constructing three log buildings to house the president's family and the students. Wheelock also had to launch a village (tavern, store, blacksmith, printer, barber, tailor, carpenter, etc.) to support the College. Yet despite its rustic environs, the new college quickly established a museum, library and "philosophical apparatus."

The museum and library can be quickly described. Already in 1772, a young missionary who had taught at a charity school also run by Wheelock wrote from western Pennsylvania that he had "collected a few curious Elephants Bones found [at Big Bone Lick in Ohio] for the young Museum at Dartmouth." A mastodon molar, still in the College collection, is thought to be from materials shipped from Ohio. Donated fossils, minerals, exotic zoological and botanical specimens and Native American artifacts continued to arrive in Hanover. By 1790 the "museum" had its own 12 x 40-foot room in the College's largest building. An early inventory, from circa 1810, lists more than 400 items, including "iron ore from Tennessee...a piece of Lava with which they pave the streets of Naples...petrified worm...a pearl Shell, sent from the East Indies...two fragments of the French National Prison, the Bastile...Indian Sachem's Cap...The Zebra...Hindu Coins...[and] A Warpipe, from the Chippewas on Lake Superior." "A pleasing collection of natural curiosities," noted a visitor in 1811.[153]

The earliest inventory of the college library (1775) lists more than 300 volumes, mostly from President Wheelock's personal library. Stored in the house of Bezaleel Woodward, the College's first tutor of mathematics and natural philosophy (and Wheelock's son-in-law), the early library consisted primarily of religious texts–sermons, Bibles and commentaries, church history, and natural theology. A few classics of eighteenth-century natural philosophy could also be found–Henry Baker's *Employment for the Microscope* (1764), Benjamin Martin's *Philosophia Britannica, a New and Comprehensive System of the Newtonian Philosophy* (1759) and George Berkeley's *New Theory of Vision* (1709). Dartmouth students

soon found this library unsatisfactory. During the 1780s, two student literary societies established their own libraries. By 1828, their combined holdings of about 8,000 volumes would dwarf the 3,500 volumes of the college's library. However, in these early decades, all the libraries, and the college museum, remained locked and inaccessible most of the time. The extent to which either actively served students or faculty remains difficult to assess.[154]

Dartmouth's initial acquisition of philosophical apparatus, equally serendipitous, appears to have been prompted by a vitriolic exchange about the "infant College" that appeared in Boston newspapers during the fall of 1771. In September, President Wheelock had sent the *Boston Evening Post* a glowing account of the College's first commencement. Several weeks later, an anonymous correspondent, signing himself Philosyllogôn, called upon the people of New England to support the new college with their prayers, financial gifts, and young men as students. Fulsomely praising the moral virtue, Christian piety, and manly nobility of Dartmouth's students, Philosyllogôn ("friend of colleges" or "lover of collections" according to a subsequent correspondent) did allow that the institution faced one shortcoming:

> They study the Arts & Sciences as at other Colleges; and if for want of a compleat Apparatus, they cannot so much excel in their experimental knowledge of the material Heavens; it is more than compensated by their experimental Knowledge of Jesus Christ, which many of them hopefully obtain.

A sarcastic reply from "Plain Dealer" reasoned that the lack of apparatus must be "no impediment to their experimental knowledge of the material earth, and consequently, of natural philosophy." Another anonymous correspondent lambasted Philosyllogôn for "spiritual pride" in claiming that the lack of apparatus for teaching the "principles of Astronomy" was more

208

than compensated for by an "experimental Knowledge of Jesus Christ. Is not human nature the same at Dartmouth as elsewhere? Or are the Governors of that Society endowed with any special power from on high of converting their pupils from sin to holiness?" Suddenly, an innocent but pedantic attempt by Philosyllogôn to support the new college had turned ugly. For Boston's newspaper readers, Dartmouth was becoming a laughingstock.[155]

New Hampshire Governor John Wentworth, Wheelock's most prominent supporter, moved quickly to repair the damage. In May 1772, John Phillips, a former New Hampshire schoolmaster turned businessman and educational philanthropist (Phillips Andover Academy and Phillips Exeter Academy), offered £175 "to assist in procuring a Philosophical Apparatus for this College." Phillips had contributed to Wheelock's earlier educational ventures and would become a major benefactor and trustee of Dartmouth. Later that summer, Wentworth invited a number of prominent guests to join him at the college's second commencement. Along with Phillips and members of the provincial assembly came Samuel Holland, British Surveyor General of the Northern District of North America. The next year Holland commissioned Heath & Wing of London to make a handsome horizontal sundial for the college (see entry), now the earliest dated piece in the King Collection. "A mathematical and philosophical apparatus" and a library would soon be available, announced Wheelock in his ongoing *Narrative of the College*, printed late in 1772. The embarrassment provoked by Philosyllogôn had been mollified.[156]

Yet converting the Phillips gift into apparatus proved difficult. Lacking an agent to buy instruments in Europe, the trustees in 1773 sent the money to Governor Wentworth, whose "friends in England" were to advise on the purchases. By the next year Wentworth reported that the London maker, Jesse Ramsden, "incomparably the best hand in Europe," had been commissioned to build Dartmouth's instruments.[157] The outbreak of the Revolutionary War disrupted communication, however, and Ramsden's apparatus eventually found its way to Russia. Wentworth himself was forced to return to Great Britain and Phillips' £175 seemed to have vanished.

After the war, Wheelock's son John, who had become president of the college in 1779 at his father's death, traveled to Europe to raise funds for the struggling school. In England, he tried to locate the Phillips gift. Governor Wentworth had left London for Nova Scotia, however, and the whereabouts of the funds remained murky. Wheelock did manage to extract promises from Paul Wentworth, the governor's cousin and planter with land in the West Indies, and his friend, a Dr. William Rose of Chiswick, to provide the College with some apparatus. In 1785, their donation finally arrived in Hanover. It included a set of mechanical powers (see entry), celestial and terrestrial globes, air pump, orrery (see entry), "standing telescope with achromatic glasses," thermometer, barometer, and an electrical apparatus. These instruments joined several miscellaneous items that had earlier been secured: the Holland sundial of 1773; two eighteen-inch globes and a solar microscope that a visitor had described in 1774; and a Hadley's quadrant by Spencer, Browning & Rust of London.[158] The latter had been donated in 1783 by John Hurd, a former associate of Wentworth's who had lived north of Hanover and had deeply involved himself in land deals before the Revolutionary War.[159] These gifts from Governor Wentworth's friends provided Dartmouth's first philosophical apparatus.

When Yale's president visited Dartmouth in 1787, he described the collection as follows:

209

Air pump double barreld & doub. plated with Receivers, &c.; 3 setts of Globes 14 Inch. 16 & 18; Electrical Appa[ratus] (small); Planetarium [i.e., orrery]; 2 setts of Mechanical Powers; Acromatic Telescope, 30 Inc[hes], 3 1/2 or 4 Inc[ches] Diam., with brass Drawers; Baker's Microscope [i.e., solar microscope[160]]; Sextant; Godfry's Sextant [i.e., Hadley's quadrant].[161]

In contrast to the 77 handsome items Harvard purchased from London makers in 1764 to replace its collection lost in a fire, or the 80 instruments Union College acquired in 1796 for £200 from London, Dartmouth's initial philosophical apparatus was relatively modest. Nonetheless, its fifteen or so objects, plus the sundial mounted in President Wheelock's garden, covered most of the major fields of late-eighteenth-century natural philosophy–pneumatics, electricity, mechanics, astronomy, microscopy–and surveying. At the end of the eighteenth century, similar sets of apparatus could be found at many colleges, universities and even pre-collegiate academies of Europe and North America.[162]

Available sources provide few details about how these apparatus were used. The earliest descriptions of Dartmouth's curriculum, from around 1800, place the natural sciences and mathematics in the second and third years. Algebra, geometry, mensuration, surveying and navigation are prescribed for all sophomores, natural philosophy and astronomy for juniors, a mandatory pattern, widely repeated at other American colleges, that would remain essentially unchanged until elective courses were introduced in the 1880s. The early curriculum prescribed natural philosophy textbooks by leading London popular lecturers and instrument makers such as Benjamin Martin, James Ferguson, George Adams and William Jones, each of whom had accompanied his lectures with theatrical demonstrations. Presumably, Dartmouth's professor of "mathematics and natural philosophy," a position established in 1782, employed apparatus for lecture demonstrations. By the 1820s, a faculty report on instruction referred to "experiments to be exhibited" in natural philosophy and chemistry, demonstrations that could not be conducted in the usual recitation rooms where the other subjects were taught.[163]

In 1790, the trustees reserved for the philosophical apparatus a small room on the third floor of Dartmouth Hall, the college's first major building and at 150 x 50 feet reputedly the largest of its kind in New England. They also instructed the president to draw up "suitable and proper regulations" governing use of the apparatus (no such rules have been found). In 1801, the trustees approved an expenditure of $150 for the purchase of additional instruments and repairs, using funds from the original Phillips gift that finally had been retrieved from former Governor Wentworth's sequestered New Hampshire estate. Presumably, a larger achromatic telescope (see entry) was purchased at this time.[164] And at some point before 1810, the College must have acquired a surveyor's compass (see entry, page 28), for President Wheelock in that year measured the magnetic deviation at Hanover for a national survey organized by a United States congressional committee.[165]

Thus by early in the nineteenth century, the philosophical apparatus, like the museum and libraries, had become a regular feature of the College, prominently displayed in its central building. In 1806, a Pennsylvania professor issued the first American edition of a four-volume *Lectures on Natural and Experimental Philosophy*, by the London instrument makers George Adams and William Jones. The subscription list for this textbook, filled with plates lavishly illustrating the apparatus of contemporary natural philosophy, includes forty-eight students from Dartmouth, two from Princeton, plus 140 other men from across the new nation. Dartmouth students, who

Chandler School surveying students, 1884

would recite from Adams' *Lectures* for the next decade, apparently appreciated the theatrical demonstration of the college's apparatus. A former student of the Seminary at Quebec City, which had one of the largest cabinets in North America, lauded experiments he had witnessed in 1825 at "Dartmouth University."[166]

Only a few of the eighteenth-century instruments (orrery, telescope, octant, mechanical powers, sundial) have survived to the present. The early air pump, globes, electrical machine, and microscope have disappeared.

Surveying (mathematical) instruments

Courses in surveying, mensuration and navigation had been required of all Dartmouth students since the founding of the College. By 1839, the college catalogue indicated that "surveying and leveling" would be taught "with the use of instruments and practice in the field," a phrase that remained in the catalogues through the 1880s when an elective curriculum was introduced. Writing his parents in 1841, a Dartmouth student indicated that the professor of mathematics would "accompany us with his instruments into the field in order to give us instruction concerning their use and also to illustrate more fully the Science." Other student notes from these years indicate that students themselves surveyed plots with chains, transits, quadrants and surveyor's compasses.[167]

Such courses provide the earliest known examples in Dartmouth's curriculum where all students systematically gained hands-on experience with instruments. These practical exercises undoubtedly explain the rich

211

collection of nineteenth-century surveying apparatus—circumferentors, graphometers, theodolites, transits, Wye levels, a Rochon micrometer, surveyor's chains and steel tapes—found in the King Collection. Elective courses in surveying and navigation remained in the modern curriculum through 1952 and 1963, respectively. Richard Goddard, one-time Arctic explorer and then professor of astronomy and from 1934-62 final director of the college's Shattuck Astronomical Observatory, especially emphasized navigation. Reflecting this curriculum, the King Collection retains a rich variety of early- to mid-twentieth-century sextants by English, German, French, American, and in one case, even Japanese makers.

Surveying apparatus also were used in two specialized schools founded at Dartmouth around 1850. The Chandler School of Science and Arts, established by a large bequest from a New Hampshire businessman, offered a two-year curriculum in "the practical and useful arts of life," including civil engineering, architecture and drawing, carpentry, the manufacture of machinery, materials, bookkeeping and "such other branches of knowledge as may best qualify young persons for the duties and employments of active life." Surveying, leveling and "field work" remained central for Chandler students until the 1890s, when, after years of friction over how "practical" Dartmouth should be, the School was collapsed into the College's regular curriculum. More lasting was the Thayer School of Civil Engineering, the nation's first post-graduate engineering school, which opened in 1872 and continues even today. Surveying served as the initial course for the first-year engineering student. Already by 1873, Thayer's catalogue announced that "[f]or practical operations in the various branches of surveying there is a good supply of all the necessary instruments, of the best manufacture." Surveying would remain the first

course for Thayer students until 1962. Many of the surveying pieces now in the King Collection came from the Thayer School.[168]

Chemical apparatus

In 1798, Dartmouth's trustees approved the establishment of a separate medical school at the college and formulated its regulations in detail. The lectures in chemistry and materia medica were to be "accompanied with actual experiments tending to explain & demonstrate the principles of chemistry and an exhibition of the principal medicines used in curing diseases."[169] Although medical school president Nathan Smith had difficulty finding someone to teach chemistry, the subject must have been taught for in 1805 the trustees provided Smith with $390 for chemical apparatus and books. In 1809 a medical student wrote home that "I have been employed…with 5 others in performing chemical experiments till 3 o'clock in the morning two thirds of the time since the lectures have begun." The notion that medical students would conduct unsupervised chemical experiments in the middle of the night seems farfetched. Yet by the 1790s, even undergraduates were writing commencement essays on chemical topics. Clearly, chemistry had established at least a nominal presence during the early decades of the young college.[170]

Rhetoric about a "laboratory" also emerges at this time. The 1810 inventory of the college museum lists several vials of chemicals from the "Laboratory belonging to Prof. [John] Hubbard," the professor of mathematics and natural philosophy from 1804-10. A eulogy delivered at Hubbard's death in 1810 describes him "as a philosopher, surrounded with the apparatus of science, extending his researches to the phenomena of the universe" and always eager to emphasize the moral and religious implications of the natural world.[171] In 1813, the trustees added chemistry to the prescribed

212

college curriculum for juniors, and began the practice that would continue for decades, often unhappily, of requiring the undergraduates to attend the medical school lectures in chemistry. Although the undergraduates complained about being treated as second-class citizens in this situation and the faculty complained about not being paid enough for teaching both sets of students (in 1825, each junior had to pay a fee of sixty-seven cents for the chemical lectures), the arrangement did bring chemical apparatus to the College.

The earliest inventory (1815) of chemical apparatus "now in the Laboratory belonging to Dartmouth College" includes not only retorts, receivers and other glassware but also a barometer, thermometer, pyrometer, blowpipe, Webster's lamp furnace, Papin's digester (pressure cooker), Watt's air holder (pneumatic trough), a small electric machine, and Nooth's apparatus (glassware designed to prepare water containing carbon dioxide).[172] As historian Stanley Guralnick has noted, chemistry with its practical ties to medicine, industry and "the arts" was the science "most successfully pursued" in the United States before 1850. Guralnick found that chemistry instruction began in American medical schools, and in the 1820-30s gradually moved into the collegiate curriculum. Likewise, Dartmouth's initial laboratory with its dedicated space and specialized apparatus was located in the college's medical building, opened in 1811; by the second decade of the nineteenth century, chemistry also had become mandated for Dartmouth's undergraduate students.[173]

For most of the college's first century, the vitality of its chemistry oscillated between extremes, depending on the availability of funds for the laboratory and the quality of the instructors. For the two decades following the 1815 inventory, the trustees provided a total of only $325 for the chemical laboratory and conditions deteriorated.

In 1836, Oliver P. Hubbard, Yale A.B. 1828, was hired to teach chemistry and pharmacy at the medical school and chemistry and mineralogy at the college.[174] After his initial visit to Hanover, Hubbard described the situation for his mentor, the preeminent Yale chemist Benjamin Silliman, in whose laboratory he then was serving as an assistant:

> But to my Laboratory–Chem. Lab. it can hardly be called. The furnaces in a block of masonry are two–one for heating a sand bath alone & another for heating an iron pot (for water) 10 or 12 quarts capacity–they can be used for nothing else. There is a little hot iron stove to warm the Lab. room & a large one in the lecture room, a black furnace in the lecture room, but the chimney running up thro' the middle of the room has been cut off…and is useless. There is not a bell glass, or any article of pneumatic chemistry, no means of making oxygen, cistern of wood about 24 in[ches] by 14 x 8, caulked with rags & cement & will not hold water. Reagents & chemicals decomposed, or corks left out & gone, or in exceedingly small quantity. Pestles without mortars, mercurial cistern no mercury or glasses but one, flat long narrow neck, no [test] tubes. Table lamps without wicks, only half of any thing & in general the utmost destitution of every useful object & only the ruins of what once was–tis like a ship too poor to be repaired, on its last voyage, abandoned at sea.

Moved by Hubbard's plight, the trustees agreed to provide $700 for chemical apparatus and furniture. By working through the summer of 1836 Hubbard apparently managed to repair the "ship" so that he could begin teaching that fall. Over the next decade, Hubbard received about $100 annually for "laboratory expenses," i.e., student assistants, a stove, charcoal and firewood, cleaning, glassware, reagents, a thermometer, and several galvanic batteries. Occasionally he purchased other unspecified apparatus from the Boston firm of Joseph M. Wightman.[175]

Judging from lecture notes, it would appear as if Hubbard's expenditures mostly supported his lecture demonstrations.[176] In 1849, he again beseeched the trustees for additional support, complaining that for thirteen years no new chemical apparatus had been purchased. Hubbard himself had been obliged to pay for a student assistant. Students now volunteered, several years in advance, to assist him with the lectures. And Hubbard suggested that all students should have the opportunity for practical work at the bench:

> [T]he means for pursuing analysis by students—& more, by myself are too imperfect, they are not capable of determining with precision the composition of a simple body or even the proper weight of any element.... it is impossible to meet the reasonable expectations of the public or individuals when applying for the analysis of minerals & the determination of the value of ores, a demand which ought to be met as well as one for surveying a field.[177]

He received $150 for apparatus and reagents, but in the 1850s, his annual spending dropped to about $50 per annum. The college catalogue for 1850-51 announced that the "Chemical Laboratory is amply furnished with Apparatus and Chemicals for the illustration of the Lectures in that Department," mentioning nothing about practical exercises for students. Hubbard's true interest, it would appear, was in geology and mineralogy, fields in which he published original researches. Although chemistry had been a required part of the college curriculum since 1813, the chemical apparatus and laboratory remained rather rudimentary during most of Dartmouth's first century. Professors performed demonstrations; students watched.[178]

Not until 1871 would a new building be erected, with an entire floor of nine rooms devoted to chemical laboratories and lecture halls. Impetus for its development came not from the college but from the

Chemistry laboratory in Culver Hall, 1876

214

newly established New Hampshire College of Agriculture and the Mechanic Arts, an institution that in 1893 would move to Durham as the University of New Hampshire. Jointly funded by the state legislature, a local farmer who advocated scientific agriculture, and the College, Culver Hall provided bench space for 30 students of practical chemistry as well as facilities for demonstrations accompanying chemical lectures. Outfitted with steam heat, gas lighting, running water and more than $1,600 of chemical apparatus and furniture, the new building contained "as fine a laboratory as any in the country," according to its local admirers. Culver Hall was the first building Dartmouth erected to house laboratories in which large numbers of College students could conduct practical exercises. The College's catalogue for 1871-72 announces the new laboratory as offering students "ample opportunity for experimental studies and manipulations, under the direction of the Professor and Assistant Instructor in that department." Optional exercises in "practical chemistry" for the senior class enter the curriculum at this date.[179]

By the 1890s when the agricultural college departed, Culver Hall had become overcrowded and the professor of chemistry called for its expansion. A $5,000 renovation in 1905 allocated the entire building to chemistry, created bench spaces for 168 students, each provided with water, gas, and air under pressure or vacuum, and carved out a special room for spectroscopy. Yet overcrowding continued and by 1913, the faculty for the first time complained about lack of private laboratory space for their own researches. Finally in 1921, a new Steele

Students in the chemistry laboratory, spring 1887

Chemistry Building was completed, containing both large and specialized student laboratories and "ample provision for offices and private laboratories for instructors and for student research." With equipment and furnishings, Steele cost a total of $475,000, making it the most expensive building constructed to date at the College. Nearly all of the eleven courses in chemistry offered by that time required practical exercises. Within several years, over 400 students a year were taking laboratory work in chemistry, requiring an annual expenditure of $4,000 for reagents and $1,500 for other materials (6,000 test tubes, 2,000 cork stoppers, 2,000 feet of rubber tubing, 4,500 feet of glass tubing, 50,000 sheets of filter paper and 115,000 strips of litmus paper). Nonexpendable apparatus was not mentioned. Nonetheless, chemistry by the 1920s clearly had become a major conduit for the influx of scientific apparatus to the college.[180]

Teaching chemistry brought apparatus to Dartmouth; research in chemistry did not, at least until well into the twentieth century. Chemistry was the first of the scientific disciplines in which Dartmouth faculty published original research. James Freeman Dana, Harvard A.B. 1813 and M.D. 1817, taught chemistry and mineralogy at Dartmouth from 1820-27. Before coming to Dartmouth, he had worked as an assistant to Harvard's professor of chemistry and in 1815 had journeyed to England to purchase chemical apparatus for that college. In the early 1820s, Dana was among the first Americans to investigate the newly discovered phenomena of electromagnetism; he was also the first Dartmouth professor to author a textbook for the sciences. His papers in the *American Journal of Science and the Arts* describe experiments with Leyden jars, a battery of 200 six-inch square plates, glass tubes, coils of various types of wire, and even a prism. He also published a design for an improved air pump, admit-

ting however that he lacked "the means for making a practical trial of the principle." Apparently, Dana personally purchased and owned the apparatus he used; none of it remains in the King Collection today.[181]

Over the next century, however, most of Dartmouth's chemists published rather less research than had Dana. Hubbard (taught 1836-66) did publish regularly in Silliman's *Journal*; yet his articles treat the mineralogy and geology of New Hampshire, not chemistry.[182] Long-time, highly esteemed chemistry professors such as Edwin J. Bartlett (taught 1878-1920), Leon B. Richardson (1900-48), and Andrew J. Scarlett (1911-57) authored general chemistry textbooks with laboratory exercises, but very few research articles.[183] Not until the second half of the twentieth century would Dartmouth chemists engage more systematically in original research.

Despite the long tradition of chemistry at the College, the King Collection contains relatively few chemical artifacts, especially from the College's first century. Only one item on the 1815 inventory remains extant. The apparatus of Culver Hall (the building was demolished in 1929) have largely disappeared, except for several analytical balances and thermometers from the late nineteenth century. None of the electronic analytical equipment, developed in the 1930s during chemistry's "instrument revolution," remains today.[184] Although chemistry brought myriads of instruments, glassware and reagents to the college, the rates of discarding and acquiring by Dartmouth's chemists have remained roughly equal. Apart from some twentieth-century Bunsen burners, balances, and thermometers, the King Collection's chemical holdings are rather thin.

We might explain this relative lack of chemical apparatus in several ways. Unlike the instruments of natural philosophy, many chemical artifacts are inherently disposable; they are consumed in the laboratory. Re-

216

INDIVIDUAL LABORATORY APPARATUS

1 burner	1 flask, Erlemeyer, 200 cc.
1 wingtop, for burner	2 funnels, 65 mm.
1 burette — 25 cc. (Mohr's)	12 test tubes, 13 cm. × 1.5 cm.
1 burette clamp	2 test tubes, hard glass, 15 cm.
1 pinch clamp	× 1.5 cm.
1 cylinder — graduated, 50 cc.	2 test tubes, 15 cm. × 2.5 cm.
1 deflagrating spoon	1 test tube holder
4 cylinders for gas collection,	1 test tube rack
250 cc., or 4 wide-mouthed	4 glass covers for cylinders
bottles, 250 cc.	1 thistle tube
1 bottle — 2 liter	1 retort, 250 cc.
1 bottle, wide-mouthed, 90 cc.	1 watch glass, 5 cm.
2 beakers, 100 cc.	4 watch glasses, 8 cm.
2 beakers, 250 cc.	1 ring-stand and 2 rings
1 beaker, 400 cc., pyrex	1 thermometer, Centigrade (− 10°
1 beaker, 600 cc., pyrex	to 110°)
1 crucible and cover	1 pneumatic trough (pyrex tray)
3 evaporating dishes, porcelain, #0	1 calcium chloride tube
1 flask, Florence, 200 cc.	1 porcelain boat
1 flask, Florence, 500 cc.	

NON-RETURNABLE APPARATUS

1 towel	1 tube, hard glass, 30 cm. × 1 cm.
1 pipe-stem triangle	4 ft. tubing, glass, 6 mm.
1 box matches	2 ft. tubing, rubber, 6 mm.
2 vials litmus paper	1 glass rod, 30 cm.
1 package filter paper, 11 cm.	1 rubber stopper, two-holed #6
1 taper, wax	1 " " " #5
1 file, triangular	1 " " one-holed #5
1 wire gauze	1 " " " " #4
1 sponge	2 wood splints
1 platinum wire	

GENERAL APPARATUS

Lens	Cylinder, graduated, 500 cc.
Blast lamp	Boyle's law apparatus. Fig. 8
Barometer	Thermometer, grad. in 1/10°
Hydrometers	Mortar and pestle
Balance and weights	Cork borers
Conductivity apparatus. Fig. 15	Blue celluloid screen
Hygrometer. Wet and dry bulb	Hand spectroscope

with oxides and corrosion that are difficult to remove and eventually degrade the function of the apparatus. We might also ask whether chemists attempt to preserve their "old" apparatus as systematically as do professors of natural philosophy or physics. In the growing number of university instrument collections that have been formalized, apparatus of chemistry generally seems less numerous than that of physics. Is physical apparatus "retired" to storage areas when superseded by new designs, whereas chemical apparatus is discarded after it "breaks" or "wears out?"

A list of apparatus from a 1929 laboratory manual (left) for the college's introductory chemistry course might confirm such speculation. Most of its items are glassware or porcelain. Other materials such as tubing, wire, litmus and filter paper, and matches are explicitly described as consumable. Only a few more robust items are listed (thermometer, barometer, hygrometer, hydrometer, balance and weights, blast lamp, and hand spectroscope), types of apparatus that now can be found in the King Collection.[185] In any case, despite the early establishment of a chemical lab at Dartmouth, chemical instruments are not very numerous in the King Collection; perhaps 100 of its 3,000 artifacts are chemical.

agents, of course, get used or are discarded as chemicals of improved grades of purity become available. Glass, the constituent of so many containers and instruments for the chemical laboratory, is less robust than the brass, iron and wood of physical instruments. Even borosilicate glasses, invented in the "Pyrex revolution" of the 1920s and prized for their enhanced resistance to thermal and physical shock, can chip during laboratory operations. And unlike instruments of brass or wood, "broken" glassware usually is easier to replace rather than repair. Likewise, spills and even normal laboratory usage can encrust metal apparatus

From philosophical apparatus to instruments of physics

The bulk of the King Collection came to the college as philosophical apparatus in the nineteenth century and as instruments of physics in the twentieth. As with chemistry, the pattern of purchases rose and fell with the interests of the instructors and the availability of funds. The larger volume of purchases also reveals interesting shifts in the international market for scientific instruments. Initially, most of Dartmouth's apparatus came from London. By the mid-nineteenth century, French makers predominated. German instruments started arriving in the 1870s. Some nineteenth-century items came from John Prince of Salem, Massachusetts, and the well-known Boston makers, Daniel Davis, Jr., N.B. Chamberlain, and E.S. Ritchie. After 1900, Dartmouth's physicists increasingly turned to American makers. This pattern undoubtedly has been repeated at many North American colleges and universities.

The "Dartmouth College Case"

The first significant expansion of Dartmouth's philosophical apparatus in the nineteenth century involved the celebrated American maker, John Prince, and the notorious "Dartmouth College Case." In 1815, the trustees fired President John Wheelock after a long series of disagreements over who should control the college church, revivalism and discipline among the students, and new faculty appointments. Turning to a group of Republican supporters in the New Hampshire legislature, Wheelock tried to wrest control of the College from the trustees, abrogate the original charter, and create a new "Dartmouth University." The trustees challenged the action in court, insisting on the validity of the charter and the College's right to exist as a private institution free from interference by the state. Daniel Webster, Dartmouth 1801, argued the case before the U.S. Supreme Court, which in February 1819 decided in favor of the College in a classic ruling that strengthened the contract clause of the Constitution and gave the College a lasting image of itself as a righteous David facing down an evil Goliath. Recent historians have interpreted this episode as the victory of emotional,

even irrational religious sectarians over Wheelock's more "liberal" religiosity.[186]

Whatever their sundry disagreements, both sides desired philosophical apparatus. From March 1817 until the spring of 1819, the university and its new faculty took over the principal college building, Dartmouth Hall, which housed the library, museum and philosophical apparatus.[187] The College, which retained most of the students, had to conduct its classes in private buildings in the town of Hanover, initially without an apparatus.

Ebenezer Adams, Dartmouth 1791, named professor of Latin and Greek in 1809 and professor of mathematics and natural philosophy the next year, had early joined the "college side" against Wheelock. He had also sought to enlarge the apparatus. In 1811, the trustees voted to ask the state legislatures of New Hampshire and Massachusetts for permission to raise $30,000 by lottery to increase the College's philosophical apparatus, library, and buildings. The New Hampshire body prohibited such ventures and this scheme went nowhere. In 1813, the trustees authorized $1,000 for a "small philosophical apparatus," provided that the funds could be borrowed. Adams personally loaned the college $540 so that he could purchase apparatus from John Prince of Salem, Massachusetts. According to later reports, the Prince purchase included an electric machine (see entry) and battery of Leyden jars. Given the prices Prince charged other universities for instruments, Adams must have acquired considerably more apparatus for his $540.[188]

Yet in the spring of 1817 these Prince purchases went to the university. Adams thus sent his colleague and fellow College defender, medical professor Reuben D. Mussey, to Boston and Salem to procure replacements, including an electric machine ($24). Given the precarious state of the College's finances, Adams also arranged for

218

Prince to loan him an air pump ($10), a set of convex, concave and cylindrical mirrors, three lenses and a perspective glass consisting of a mirror and lens (for a total of $6). Later in 1818, Adams purchased an "opake and transparent solar microscope" from Prince for $70. Eager to advertise its ongoing viability, the College placed a circular in New Hampshire's newspapers:

> The Trustees...are happy to announce to the publick, that, through the munificence of friends in Boston, the Professor of Mathematicks and Philosophy will be furnished, the ensuing season, with a good electrical apparatus, an Air Pump with accompanying instruments, an elegant Telescope, a new Solar Microscope, and a common Microscope, which, together with other articles in his possession, will enable him to exhibit the most important experiments in all the branches of Natural Philosophy.

In November 1819, after the trustees sued the president of the former university for the return of the library and apparatus, Adams sent a microscope and telescope back to Prince. The fate of the other Prince "rental" items is not revealed in the extant documents. Professor Adams, who as a schoolteacher in 1797 had "revised and corrected" a popular arithmetic text but otherwise would publish nothing in natural philosophy, nonetheless revealed himself to be a keen advocate of experimental lecture demonstrations. Between 1814 and 1822, he spent about $725 on apparatus. The College catalogue from 1824 on describes Adams as offering "experimental lectures." A member of the class of 1827 later remembered the apparatus during Professor Adams' time as "small and inexpensive."[189] Yet under most inauspicious circumstances, Adams had managed to procure apparatus from America's most prominent maker and thereby to signal the vitality of his besieged college.

Joining international science
For Adams' successor, philosophical apparatus became a means to give Dartmouth a competitive edge over other American colleges and to enter an international world of scientific research. Initially trained as a carpenter in the neighboring town of Lebanon, New Hampshire, Ira Young occasionally worked for the College and was given access to the apparatus by Professor Adams. Desiring more, Young at the age of 21 prepared for college, entered Dartmouth with the class of 1828, and returned in 1830 as a tutor. Recognizing Young's abilities, Professor Adams upon retiring in 1833 urged the trustees to appoint the former carpenter to his professorship of mathematics and natural philosophy. In 1838, Young became professor of natural philosophy and astronomy (only the second American professorship for astronomy). Young also married Adams' daughter, a very common career move for nineteenth–century professors. Yet Young's career at Dartmouth would be anything but common. Until his death in 1858, the local lad would take steps to craft a vision for a scientific infrastructure that could support research as well as teaching, bring French and German instruments to Dartmouth, and secure a major endowment that enabled Dartmouth to build an astronomical observatory, its first building dedicated to science.[190]

Young honed his vision by comparing Dartmouth with other educational institutions. Soon after assuming his professorship, he spent a month visiting apparatus collections and instrument makers in Amherst, New Haven, Hartford, New York, Philadelphia, Princeton, Providence and Boston (the first trade catalogue of instruments offered by an American maker would not be published until 1834). On this trip, Young purchased several big-ticket items: an alt-azimuth theodolite ($200), Atwood's machine for demonstrating mechanical advantage ($85), and armillary sphere ($15). He also bought meteorological apparatus (thermometers and barometers) and with his father-in-law, Ebenezer Adams, began keeping a daily meteorological register. These efforts, "important," Young claimed, "not only to the general interests of science but to the reputation of the College," represented the first systematic use of instruments at Dartmouth for research rather than lecture demonstration. Yet Young's initial purchases also included low-cost items for demonstrating amusing optical effects: a perspective glass, multiplying mirror, Claude Lorraine glasses, and phenakistocope (rotating disk with slits that display "moving" pictures, invented by Joseph Plateau in 1830). In all, he spent $370 for apparatus in 1834, plus $60 for his travel expenses.[191]

Young also constructed many articles for "practical illustrations" in college workshops. In 1836 he worked with "village mechanics" to make a "Hero's fountain, spiral pump, model of Barker's Mill, electrophorus, torsion balance, apparatus for polarization of light, telescopic kaleidoscope, solar kaleidoscope, [and] several smaller illustrations in electricity and astronomy." Like his predecessor, Young is described in the College catalogue as giving "experimental lectures on natural philosophy." Indeed, his showmanship in the classroom began to gain him a following among some of the students.[192]

Young kept pushing natural philosophy toward the center of Dartmouth's physical and cultural landscape. In 1839-40, he oversaw the construction of a major new building for the College, erected in response to expanding enrollments in the 1830s.[193] Reed Hall provided space for the College's libraries, museums, and philosophical apparatus plus a large, sloped auditorium for lecture demonstrations (see entry Housing Physics). Designed by his brother, Ammi, and constructed by another brother, Reed Hall was clearly a Young family project. Its success enhanced Young's local influence. In 1841 the trustees announced an ambitious $50,000 subscription plan, seeking to endow several professorships, increase the library and the philosophical, astronomical and chemical apparatus, and to erect an astronomical observatory (the first time such a project had been mentioned). By 1845, as the subscription threatened not to reach its (now reduced) goal of $30,000, a prosperous Boston merchant and cotton manufacturer, Samuel Appleton, suddenly stepped forward to contribute $10,000 for a professorship in natural philosophy (not one of the professorships specified in 1841!). Extant sources do not indicate why Appleton, who had no earlier connection to the College, decided to support natural philosophy. Undoubtedly, Ira Young played some role in his decision. In any case, the Appleton Professorship was the most handsomely endowed chair the College had received to date. And after his death in 1853, Appleton's estate added another $15,000 to the endowment.[194]

Despite Appleton's generosity, many pledges of the 1841-45 subscription failed to materialize and the next year a Faculty Committee on the Library and Philosophical Apparatus implored the trustees for additional support. Signed by Young and Charles Haddock, the College librarian and philosophy professor, the report outlines a new vision for science at Dartmouth, based on a rhetoric of competition. Asking for $10,000 for the library, it argues that "no college of equal reputation...is so imperfectly provided with books adapted to the wants of the professors." To enlarge the philosophical apparatus, $6,600 is requested. "What we have cost, originally, about $2,300 [Young later revised this figure down to $1,100]; and is estimated to be now worth about $1,300. Amherst College, Wesleyan University, Williams, Princeton, & the Western Reserve Colleges have expended each from six to ten thousand dollars for appara-

220

tus; and but two of them have furnished an astronomical observatory."[195] Most of Dartmouth's apparatus had been purchased between 1812 and 1819, "when the cost of instruments was unusually high, and the articles purchased were many of them of an inferior quality, and most of them are now either worn out or broken or unsuited to the present state of science…so that the whole collection is in reality of little value." Such talk of obsolescence would infuse equipment requests by later generations of newly appointed Dartmouth scientists. Young's argument is slightly ironic, since he himself had since 1834 purchased hundreds of dollars of apparatus.

Young envisioned a coherent collection, comprised of substantial, relatively expensive items that should serve the College for at least fifty years. "Showy instruments, intended chiefly for display, or amusement, are not wanted, but a good, working apparatus, sufficient for making valuable observations and illustrating the important principles of science." Young requested an astronomical observatory, including telescope, transit circle, astronomical clock and comet seeker ($4,200); magnetic observatory ($700); meteorological instruments ($300); engineering instruments, including theodolites, level and circle ($700); and a compound microscope, solar microscope, polarizing apparatus and heliostat ($700). To justify this vision, Young appealed to a range of goals. The "present condition of our country" requires civil engineers and surveyors; possessing such instruments would attract students to Dartmouth who now seek an education elsewhere and would prepare them for practical careers. High-quality microscopes serve "actual investigation" and introduce students to a "world of wonders" that will stimulate their minds to "healthful action." A magnetic observatory, "like the one already in use at Williams College," would employ students "in their leisure

hours in making observations of a truly valuable character." Although he allowed that "a few articles for illustrating the other departments of natural philosophy are also highly desirable," the former carpenter's vision emphasized students directly engaged in making observations to produce useful scientific or practical knowledge.

Young thought big. His proposal called for spending three times the total outlay made for natural philosophy since the founding of the college. Indeed, with the College's 1845-46 budget totaling only $12,000, the faculty request for a one-time expenditure of $16,600 was unprecedented. "These sums, we are aware, appear large; they are large in comparison with what has been done, at any former period, for the same objects. We suppose, however, that, in the present state of education in New England, we can hardly maintain our position as a literary institution, without some extraordinary enlargement of our means & instruments." Young and the faculty worried about remaining competitive; the trustees, however, worried about declining enrollments and falling income. They rejected any funds for the library but did allow $2,300 for philosophical apparatus "from money yet to be collected" from the 1841-45 subscription. They approved Young's request to visit other institutions, "in order to a judicious selection of apparatus" and to gain information about other astronomical observatories. And they accepted Young's entrepreneurial suggestion that, along the way, he act as agent in collecting payments now due on the 1841-45 subscription.[196]

Young chose to begin with an astronomical observatory, the largest item on his wish list (see entry for Shattuck Observatory). After consulting with other American astronomers who had recently acquired telescopes by the Munich maker, Georg Merz, Young ordered a 6.4-inch refractor from that firm, spending the full $2,300 provided by

Ira Young's request to the trustees, 1846

the trustees. Yet they refused to fund his request of $5,800 for a permanent observatory building, allowing Young only $300 to erect a temporary shed for the telescope in his garden. Once again, however, the entrepreneurial professor proved unstoppable. In George C. Shattuck, Dartmouth 1804, a wealthy Boston physician and old family friend, Young found a patron willing to support his comprehensive vision for science at the College.

Earlier in 1821, Shattuck had provided Professor Adams with $50 to buy natural philosophical books from Europe. In 1852, he contributed another $1,000 for this purpose. Young immediately entered into correspondence with the Boston physician, describing the European periodicals he wanted as "indispensable works of reference" and worrying about the expense of importing books. To reduce costs, Young purchased nearly 150 duplicates from Harvard libraries, acquiring technically demanding French and German books and periodicals filled with the latest European research, and classic Latin texts from the seventeenth and eighteenth centuries. Unlike the rudimentary textbooks or popular treatises, all in English, hitherto available in Dartmouth's libraries, Young's Harvard purchase signaled an intent to engage with contemporary European science from London to St. Petersburg.[197]

Young reiterated his 1846 vision in subsequent letters to Shattuck, comparing Dartmouth unfavorably to what other American colleges were providing for science. Requesting again an astronomical observatory ($6,000), surveying and meteorological equipment ($1,500), Young now outlined his needs for demonstration apparatus to show the laws of rotary motion, per-

222

chiefly for display, or amusement, are not wanted, but a a good, working Apparatus, sufficient for making valuable observations and illustrating the important principles of science.

Such instruments must, of course, be comparatively expensive. They cannot be of substantial character and permanent value without being costly.

1. A Telescope of this kind, ~~would~~ properly mounted and furnished, fit to be the basis of a small but valuable Astronomical Observatory, together with the necessary accompaniments, a first rate small Transit Circle, an Astronomical clock and a Comet seeker, would cost, at least, $4,200.

2. A small magnetic Observatory with good instruments, would be about $700.

3. A first rate Compound Microscope, together with a solar Microscope, a Polarising Apparatus and Heliostat, $700.

4. ~~A~~ First rate instruments, in the department of engineering, &c. Theodolite, Level, Circle, $700.

5. Good Meteorological instruments, with a few additions to other departments, $300.

cussion and friction ($60), the laws of liquids and specific gravity ($165), the laws of pneumatics ($50), frictional electricity ($50), magneto-electricity ($100), polarized light ($120), lenses, mirrors and catoptrics ($100) and a heliostat ($160), sound and their application in the construction of musical instruments ($150 in Paris). To project images of natural objects and diagrams, he needed a new solar microscope ($130) and a magic lantern with slides ($200). Totaling the list, Young again offered a comparison, estimating that about $2,800, "when added to what has already been expended by us for philosophical apparatus, would [still] fall considerably short of what already has been expended by Amherst for similar purposes."[198]

Shattuck agreed to provide $7,000 for an observatory and apparatus plus another $1,790 for an observatory library, provided the trustees add $4,000 and that Young personally be sent to Europe to purchase the books and instruments. Calling a special meeting, the trustees agreed to borrow the required funds and accept Shattuck's conditions. In less than a decade, Young had secured two large bequests to support his vision for making Dartmouth's scientific infrastructure world-class.

Over the next years until Young's untimely death in 1858, scientific equipment poured into the Shattuck Observatory and Reed Hall. In 1853, Young and his nineteen-year-old son, Charles (Dartmouth 1853), spent five months in Europe, visiting instrument makers, book dealers, astronomical observatories, universities, and even Lord Dartmouth, grandson of the founding benefactor. Authorized to spend up to $10,000 for apparatus and books, the Youngs' purchasing trip followed a common arc established by American university professors during the first half of the nineteenth century. In 1805-6, Yale's newly appointed chemist, Benjamin Silliman, had traveled to Britain to enhance his own education at the University of Edinburgh and purchase volumes for Yale's library and some big-ticket apparatus in London. Over the next several decades, a steady stream of American natural philosophers made the transatlantic journey to buy apparatus and books, meet European scientists and instrument makers, and attend meetings.[199]

The Youngs traveled to London, Paris, Brussels, Cologne, Hanover, Berlin, Halle, Leipzig, Munich, Lucerne, Geneva, Oxford, Edinburgh and Glasgow. Ira later remarked that "[t]he marked courtesy & kindness, received from many distinguished scientific gentlemen was the more grateful, as it indicated a lively interest on their part, in the prosperity of our rising American scientific institutions." A letter of introduction from the elderly Silliman seems to have smoothed their way.[200] Despite visiting many instrument shops, they concentrated their purchases with only a few leading makers. In Paris, they acquired an acoustics apparatus from Marloye et Cie ($140) and a solar microscope, stereoscope, spherometer and polarimeter from Duboscq-Soleil ($330) (see entries). Seeking advice from astronomers across Europe, they considered buying a secondhand transit telescope and meridian circle in Rome or a new set from a German maker before deciding finally to order a more expensive, larger version from Troughton & Simms of London ($1,400). They purchased a secondhand pocket chronometer ($135), armillary sphere and some surveying instruments ($150), several barometers from John Newman in London ($135), and a comet-seeker or small telescope from Merz in Munich ($135). In all, the Youngs spent $2,500 for apparatus, $940 for books on astronomy and natural philosophy, $2,800 to stock the College library with books on other topics, including English literature, philosophy and chemistry, and $1,680 for travel expenses.[201] Nearly all of these 1853 purchases remain extant in the King Collection or Dartmouth's libraries.

Upon returning home, Ira Young oversaw the completion of the Shattuck Observatory and assembled an observatory library of well over 500 volumes in a room beneath the dome. By 1854 he had a working observatory, with a telescope that he publicly depicted as the third largest in the United States. He continued purchasing philosophical apparatus, now from American makers, especially E.S. Ritchie of Boston ($300). Yet despite his success as an institution builder, Young never published any research or textbooks based on the appara-

P.

Pamphlet Case No. 1 – contains –
  Quetelet, Meteorologie, Magnetisme Terrestre &c
    18 mo – Brussels – 1853   (French)
   "   Sur l'Electricité des nuages orageux.
     12mo – Brussels   (French)
   "   sur le principe electrostatique de Pelagi. 12mo
   "   Sur les Aurores Boreales   12mo
   "   Longitude de Bruxelles (telegraphic) 1854. 12mo
   "   Rapport sur l'Observatoire – 1854.   12mo
   "   Probable Error d'un observation du transit. 12mo
   "   Sur l'influence des Academies   12mo

Blodget – On the Summer Climate of 1853.
   12mo – Washington 1853.

Nicollet – On Meteorological Observations.
   12mo. Washington 1839.

Account of Brooke's self registering meteorological apparatus – photographic.
   4to – London 185(2?)

Pamphlet Case No. 2 contains

 Jahrbuch of the Royal Observatory at
   Munich 1839.   12mo.   (German)

Harding & Wiesen Kleine Astronomische Ephemeriden.
 1830-33 – 1830-31 &33 bis.   8vo &12mo   (German)
     Göttingen –

Charles A. Young's catalogue of the observatory library, 1857

tus he worked so tirelessly to acquire. He sent his weather observations to the Smithsonian and started to calibrate the observatory chronometer, but never initiated systematic astronomical work. In 1848 he joined the newly established American Association for the Advancement of Science, but never presented a paper at an AAAS meeting. Despite his 1853 journey, Young's involvement in science remained local, like most teachers of natural philosophy at mid-nineteenth-century American universities and colleges. In 1858, Dartmouth's President Nathan Lord eulogized Young as a Christian and consummate teacher. He did not mention that Young had attracted tens of thousands of dollars in gifts to implement his vision of international science at Dartmouth.[202]

Young's legacy is illustrated by the first major inventory of the philosophical apparatus, prepared four years after his death in 1862 at a time of financial crisis. As it had most American colleges, Northern and Southern, the Civil War hit Dartmouth hard, drying up contributions and reducing the tuition-paying student body from 360 in 1860-61 to 230 in 1864-65. Yet earnings on the Appleton endowment had grown to exceed the professorship's salary so the trustees decided to "sell" the philosophical apparatus to the Appleton Fund and to tap it for ongoing support of the apparatus. To complete this bookkeeping maneuver, the treasurer inventoried the apparatus, appraising each item's original and current value.[203]

The collection in 1862 was dominated by the apparatus of electricity, optics, pneumatics and acoustics, branches that had featured prominently in lecture demonstrations since the eighteenth century. The most expensive items listed include an induction coil by Ritchie ($150), plate electrical machine ($125), air pump by Chamberlain ($125), compound microscope by Merz

Table 1: Appleton Fund Philosophical Apparatus, 1862[204]

	Number of entries	Original cost	Current value
Materials	23	$29	$27
Glassware	17	21	19
Tools	9	11	9
Somatology	10	14	14
Mechanics	23	288	244
Hydrostatics	16	69	55
Undulations	2	5	5
Pneumatics	103	437	289
Acoustics	53	202	199
Heat	11	14	8
Optics	132	815	686
Static	81	530	422
Dynamic electricity	75	476	409
TOTALS	555	$2911	$2386

($100), and the solar microscope by Duboscq ($95). Practical apparatus favored by Young—surveying, navigation, meteorology, and the observatory apparatus—do not appear in the inventory and apparently were not considered part of the "philosophical apparatus." The inventory lists very few makers. Other than Young's 1853 purchases from Duboscq, which pepper the optics columns, the only European makers identified are W. & S. Jones of London (prism, see entry), Merz of Munich, and Carpenter and Westley of London (magic lantern and microscope). Several Boston makers appear—Ritchie, Chamberlain and Widdifield—and one item is attributed to Benjamin Pike of New York. Another dozen items are identified as "home made." The vast majority of the items were valued at less than $5 each, and probably also had been locally made.[205] At most ten percent of the items in the 1862 inventory can be unambiguously linked to instruments currently in the King Collection. Many of the surviving items appear in this book. Young's observatory remains standing to this day; most of his instruments have disappeared.

The biggest and the best

Ira Young articulated a vision for Dartmouth and international science; his son Charles would realize the vision. This episode began in 1866 when Charles Young returned to Hanover as the Appleton Professor of Natural Philosophy and Astronomy, after having taught for nine years at Western Reserve College in Ohio. During the 1853 purchasing trip, Charles had filled his diary with descriptions of workshops, machines, and the European industrial landscape. His first published paper described a printing chronograph he had invented. Clearly, Charles shared his father's practical interest in apparatus; yet unlike his father, he was also deeply committed to research. By the time he would leave Dartmouth for Princeton in 1877, Young had become one of America's best-known astrophysicists and a world-renowned authority on solar spectroscopy.

In negotiating his return to Dartmouth, Young demanded an immediate sum of $1,500 to $2,000 to "complete the outfit" of the observatory and hire a student assistant, and the assurance of continued appropriations for astronomy. The trustees approved $1,100 up front and by 1868-69 made Young the best-paid faculty member at Dartmouth ($1,800 plus $258 for Chandler teaching). They also provided, for the first time in Dartmouth's history, funds for the regular, annual purchase of apparatus and scientific books, tapping the lucrative returns of the Appleton Fund. The 1869 College catalogue proudly indicated that "large appropriations are yearly made for [the] increase and improvement" of the philosophical apparatus. After his successful spectroscopic observations of the solar eclipses in 1869 and 1870 (see spectroscope entry), Young would quickly garner an international reputation. Indeed, he is undoubtedly the most prominent faculty member of Dartmouth's first century. Young proved that scientific instruments, wielded for purposes of research, could bring fame and renown to the College. The Dartmouth student newspaper eagerly reported his every turn, his research triumphs, even his acquisition of new instruments and equipment. At the College's centennial celebrations in 1869, orators praised the new role of science in producing "engineers, machinists, men of business and professors" for the nation. During Young's tenure, scientific instruments became a significant aspect of Dartmouth's public image, not unlike their role in the 1771 Boston newspaper exchanges.[206]

Young's initial purchases, clocks, chronographs, reversing apparatus, multi-prism spectroscope (see entry), and meteorological instruments, further equipped the observatory. In 1871 after his eclipse exploits, he raised several thousand dollars by private subscription to purchase a new Clark telescope (see observatory entry), a major purchase in which the trustees played no role. He also bought a 12-inch turning lathe and tools ($100) to custom-make apparatus and accessories for the observatory. Young also enlarged the philosophical apparatus for teaching. In 1866-68, he made numerous trips to Boston, not only to work with Alvan Clark but also to purchase demonstration apparatus from the Boston firms of Ritchie, Chamberlain, and Thomas Hall. He often announced his major acquisitions—an air pump (see entry), Atwood machine, gyroscope, acoustical siren, Geissler tubes, aurora tube, induction coils, galvanometer, a Holtz electric machine, batteries, and "magnetic apparatus"—in leading American scientific journals. An 1870 article in the *Journal of the Franklin Institute*, for example, described Young's induction coil, custom-made by Ritchie for $700 and capable of generating 19-inch sparks, as the "most powerful in America."[207]

A wave of European acquisitions came in 1870-71 as Young traveled to Spain for the solar eclipse and, in a repetition of the 1853 trip with his father, also visited more than twenty instrument-making firms in London, Berlin, Leipzig and Munich. In London, he spent $960 at Henry Crouch, $460 at John Browning, and $320 at Alfred Apps; in Berlin, $610 at Schmidt & Haensch, $510 at Quilitz and Warmbrunn; in Leipzig, $320 at Stöhrer. In all, Young expended more than $4,200 for high-quality demonstration apparatus. As usual, the student newspaper rhapsodized about the instruments, yet understating their total cost:

> One of the results of Prof. Young's visit abroad last year was the acquisition to the College of a large amount of philosophical apparatus...most of the articles were made to order upon designs furnished by Prof. Young. Among other things which a very judicious investment of $500 has obtained may be mentioned the new electric light and lantern, purchased with its accompaniments at a cost of $350. An apparatus for spectrum projection; Koenig's manometric flame; Lissajous' apparatus for the combination of vibrations; a large quantity of new acoustic apparatus including a complete set of resonators and a double siren; a Malony's apparatus for measuring heat, one of the most delicate and accurate known; a fine tangent galvanometer and one of the best reflecting galvanometers that has ever been imported; a rheostat, set of resistance coils and other important electrical apparatus.
>
> In the department of optics two huge Nicol's prisms, one and a half inches in diameter by four in length and a splendid polariscope with the appurtenances necessary for the exhibition of all the phenomena of polarization should not be forgotten. The new Holz "influence" electric machine which was built in accordance with the suggestions of Poggendorff, the most eminent of German physicists is truly a remarkable instrument. This machine which will throw a spark of ten inches...is hardly equaled in America. Brown University has one with a single plate of a little larger diameter which throws a spark four or five inches farther but in construction and practical workings ours is unrivalled.[208]

Charles A. Young, 1872

Charles Young desired the biggest and best instruments he could find. The philosophical apparatus, under his tenure, had been "greatly increased by large purchases, both in this country and in Europe. In the departments of Acoustics, Optics, and Electricity, it is espe-

228

cially rich, and hardly surpassed in the United States." Yet the College catalogue also noted that these items were used for "purposes of illustration" during Young's lectures. The students themselves did not engage in experimental activities. After 1872, Young made no more major purchases. In 1877, he left Dartmouth for Princeton, lured by the offer of a new 23-inch Clark telescope and advanced astronomy students. Young's spectroscopic research had become too expensive for Dartmouth to support. "With the means at their command, [Princeton] can offer such a man as Prof. Young far better inducements than can Dartmouth at present....The reputation that has attached to the Dartmouth Observatory will be translated to Princeton," lamented the student newspaper at Young's departure.[209]

Practical exercises for physics

To a certain extent, the dismal predictions of the students would be born out in the years immediately following Young's departure. Finances and enrollments shrank in the 1880s. Dartmouth's new president, elected on the very day that Young submitted his resignation, was a clergyman who in his inaugural speech ridiculed Darwin and the excessive specialization of "modern science." A strong advocate of classical education founded on Latin and Greek, President Samuel Bartlett sought to reduce the profile of Dartmouth's Chandler Scientific School and eventually prohibited Dartmouth faculty from teaching its courses. Some alumni, students and faculty resented these moves, and Bartlett's presidency by 1881 become embroiled in controversy. Young's successor, Charles F. Emerson, openly supported the president and, in trying to make peace, increasingly became involved in administrative issues. He would not continue Young's tradition of active scientific research and expansion of the apparatus. Yet during Bartlett's and Emerson's

tenure, the teaching of natural philosophy would undergo a fundamental transformation, as students gained hands-on access to the apparatus in laboratory exercises. In 1882 natural philosophy at Dartmouth, following the path set earlier by courses in surveying and chemistry, became "practical."[210]

Young had required his students to recite from Benjamin Silliman, Jr., *Principles of Physics*, a popular American textbook of the 1860s that depicted hundreds of demonstration instruments in small wood engravings interspersed through the text. Although Silliman offered few equations, his chapters did end with quantitative problems for students to solve. When Emerson took over the natural philosophy courses, he shifted to an even more popular text by a French private physics teacher, Adolphe Ganot. *Elementary Treatise on Physics, Experimental and Applied*, a worldwide bestseller in print from 1851 through the 1920s, featured hundreds of wood engravings of apparatus and no equations. Many of Ganot's illustrations closely mirror instruments marketed by leading European makers during the second half of the nineteenth century. This textbook scripted lecture demonstrations in countless physics classrooms across Europe and America for more than two generations. Emerson used the Appleton apparatus for many lecture demonstrations, not only in his Dartmouth classes, as had Young, but also in summer schools, general lyceums and teachers' institutes across New England.[211]

Yet Emerson also introduced a significant pedagogical innovation. In 1883 the student newspaper reported that "a few days ago a freshman accosted a senior with the query, 'What room is that in Reed [Hall] where so many fellows work and they keep the gas lit so late?'" What the freshman had noticed were students conducting laboratory experiments in physics. In 1882, when Dartmouth launched elective classes for the jun-

Table 2: Dartmouth's Physics Classes, 1882-83[212]

Title	Level	Catalogue description	Textbooks
Physics 1	Jr prescribed	Elementary physics	Gage
Physics 2	Jr elective	Advanced theoretical work with experi-ments in laboratory (45 exercises)	Pickering, Maxwell, Stone
Physics 3	Jr elective	Advanced textook and laboratory work, heat and light (40 exercises)	Pickering, Stewart, Lommel
Physics 4	Sr optional	Practical work in laboratory, magnetism and electricity (44 exercises)	Pickering, Jenkins

iors and seniors, Emerson completely revised his curriculum, creating four courses in physics. The single required course he taught, as before, by lectures and demonstrations. The three elective courses, however, featured student laboratory work.[213]

The idea of moving from a fixed to an elective curriculum had prompted vigorous debate across American colleges and universities in the 1870-80s. Many scientists feared that an elective system would decrease the numbers of students in their classrooms. Yet they also worried that their teaching was becoming increasingly superficial as they tried to cover the ever burgeoning content of nineteenth-century science in a curriculum required of all students.[214] Emerson had the best of both worlds. He retained a captive audience for Physics 1 and in the other three courses offered interested students more advanced topics from textbooks by leading European physicists as well as practical laboratory work.

Emerson arranged the practical exercises around the first American laboratory manual of physics, published in 1873 by the MIT physicist Edward C. Pickering. In the place of Ganot's illustrated physics of theatrical demonstration, Pickering presented a physics of measurement and physical manipulation of instruments (his manual has almost no illustrations). He proposed a "laboratory system," specifying even the layout of the room, and exploratory learning.

> Perhaps the greatest advantage to be derived from a course of physical manipula-

tion, is the means it affords of teaching a student to think for himself. This should be encouraged by allowing him to carry out any ideas that may occur to him, and so far as possible devise and construct, with his own hands, the apparatus needed.... To aid in this work, a room adjoining the laboratory should be fitted up with a lathe and tools for working in metals and wood, as most excellent results may sometimes be attained at very small expense, by apparatus thus constructed by students. The method of conducting a Physical Laboratory, for which this book is especially designed, and which has been in daily use with entire success at the Institute, is as follows. Each experiment is assigned to a table, on which the necessary apparatus is kept, and where it is always used. A board called an indicator is hung on the wall of the room, and carries two sets of cards opposite each other, one bearing the names of the experiments, the other those of the students. When the class enters the laboratory, each member goes to the indicator, sees what experiment is assigned to him, then to the proper table where he finds the instruments required, and by the aid of the book performs the experiment.... By following this plan an instructor can readily superintend classes of about twenty at a time, and is free to pass continually from one to another, answering questions and seeing that no mistakes are made.... Moreover, the apparatus never being moved, the danger of injury or breakage is thus greatly lessened and much time is saved.[215]

By the 1880s, a "student laboratory movement" was sweeping American colleges and universities. When in 1886 Harvard began requiring its incoming students to have completed a laboratory physics course in secondary school, the movement entered American academies and high schools as well. Emerson's innovations at Dartmouth fell squarely into this transformation of American science education.[216]

Unfortunately, extant sources do not indicate how Emerson arranged his laboratory courses. For the first several years, he conducted the exercises in cramped quarters behind the large lecture room in Reed Hall (see entry Housing Physics), with the aid of an assistant, usually a recent Dartmouth graduate. In the first year of the new curriculum, 19, 12 and 24 students enrolled,

230

respectively, in Physics 2, 3 and 4 (out of a total of about sixty students each in the junior and senior classes). When in 1885 the fine arts collections moved from Reed into a new library building, Emerson took over the entire first floor of Reed for his "physical laboratories" and had added two additional courses in "original experimentation and research." By 1893, an assistant professor of physics would be appointed and the curriculum expanded into 8 courses, with Physics 1 and 2 as elementary lectures, and Physics 3 to 8 on narrower sets of topics, each including experimental exercises. By the 1890s, a method of undergraduate physics instruction had emerged, at Dartmouth and elsewhere, that would continue with few changes through most of the twentieth century.[217]

Emerson apparently made few significant purchases of instruments or laboratory furniture. In 1879, however, he did arrange for the appointment of a college carpenter with machine shop. It seems likely that Emerson purchased few commercially manufactured instruments, preferring locally made apparatus (see tangent galvanometer entry). As noted above, Pickering wanted students to construct their own apparatus. Most of the experiments in his manual require only simple apparatus that can be assembled or constructed without much skill. Only a moderate list of precision instruments, presumably to be obtained from commercial makers, are required for Pickering's student exercises: balances and weights; hydrometers, an air pump and receivers; organ bellows and pipes, tuning forks, a siren; spectroscope, prisms, lenses, microscope, ophthalmoscope; batteries and an electric machine; induction coil and Geissler tubes, galvanometers; thermometers and pyrometers; surveyor's transit and dip needle. An undated inventory (probably 1880s) of the "Apparatus of the Appleton Professorship" lists 430 items, often using names found in Pickering's manual and seldom specifying commercial makers. Like the 1862 inventory, only about ten percent of this inventory can be correlated to items currently in the King Collection. Most of the locally made apparatus from Emerson's student laboratory has disappeared. Young desired the biggest and best instruments for his own use; for Emerson, simple locally made versions, suitable for students, sufficed.[218]

A new laboratory for physics

Until the 1890s, the story of scientific instruments at Dartmouth follows patterns typical at many, if not most, American colleges and universities. But in the final decades of the nineteenth century, as the land-grant universities launched graduate programs, new universities like Johns Hopkins, Chicago and Stanford opened their doors, and older colleges like Harvard, Yale and Princeton converted themselves into universities, Dartmouth diverged from this trajectory toward the modern research university. In his 1893 inaugural address, President William Jewett Tucker outlined what he called the "historic college idea." Unlike a university, a college retains a "homogeneous character." It serves a local constituency, focuses on intellectual and religious training (rather than merely technical or cultural), and emphasizes teaching even while recognizing that "discovery stimulates teaching and that teaching necessitates discovery." Noting that the New Hampshire College of Agriculture and the Mechanic Arts had recently severed its connection with Dartmouth and that the Chandler Scientific School would soon be subsumed into the college, Tucker reiterated his view in no uncertain terms: "Dartmouth College has not been ambitious to become a university in name or in fact. The college has been, and is, and will be, ambitious to stand, with its increasing years and its enlarging strength, as the type of the historic college."

This decision would powerfully shape the college and its scientific instruments in the twentieth century.[219]

Yet Dartmouth's physics would experience one more quickening before setting into pedagogical patterns that would not change until the 1960s, when small Ph.D. programs were introduced into the natural sciences. In 1893, Professor Emerson became the first dean at Dartmouth; by 1899, he gave up the Appleton Professorship and the teaching of physics, devoting himself full time to administration. Several of his assistants would make names for themselves and Dartmouth's instruments. In 1896, Frank Austin and Edwin B. Frost, then professor of astronomy at Dartmouth, cobbled together an induction coil and Puluji tube and obtained some of the first Röntgenograms in the country (see X-ray entry), publishing the work in *Science*. Austin went on to become a professor of electrical engineering at Dartmouth's Thayer School. Frost became professor of astrophysics at the University of Chicago and then director of the Yerkes Observatory. Another Emerson assistant, Albert C. Crehore, invented a polarizing "photo-chronograph," a device that employed Nicol prisms and the Faraday Effect to make high-speed photographs of artillery shells. With this technology, he developed a method to send high-speed telegraph signals over alternating-current power lines and in 1898 left Dartmouth to manufacture this invention. Both Austin and Crehore exploited Dartmouth's apparatus for their researches; neither chose to continue their careers in Dartmouth's physics department.[220]

Despite his rejection of the university idea, President Tucker realized that the College needed up-to-date facilities if it were to retain men like Austin and Crehore. Through Frost's father, a physician and College trustee, Tucker in 1897 secured a $100,000 bequest from a local paper mill owner to build and maintain a new physical laboratory. Over the previous two decades, dozens of universities in Europe and America had erected dedicated buildings for physics. Cornell opened its laboratory in 1883, Harvard in 1884, Brown in 1891 and Amherst—still a major rival for Dartmouth—in 1892. Dartmouth's building, largely planned by Dean Emerson, would echo these earlier models, with special-purpose rooms dedicated to various branches of experimental physics (electricity, magnetism, optics, spectroscopy, photometrics, acoustics, and a "chemical kitchen") and its support (metal- and wood-working shop, equal-temperature room, dynamo room, photographic darkroom, apparatus storage, living quarters for assistants, library). Essentially, Pickering's vision, conceived for a single room, had been expanded into a four-story edifice. The building also contained a lecture hall for 200 auditors, a large two-story room for the beginners' laboratory, two recitation rooms, a laboratory for seniors (presumably independent research) and two small "private laboratories" for faculty research (see Housing Physics entry). Characteristically, the student newspaper called Wilder Laboratory "one of the handsomest physical laboratories in New England." With this building, "the Dartmouth of the future is fairly in view."[221]

Yet a laboratory without apparatus is of little use, asserted Ernest Fox Nichols in 1898. Emerson's successor, Nichols had completed a Ph.D. in physics at Cornell and had just returned from spending two years in the Berlin laboratory of Heinrich Rubens, one of the world's premier experimental physicists. Nichols' plea to President Tucker reiterates the rhetoric of Ira and Charles Young, respectively, when they had assumed their teaching posts at the College:

> I do not think it possible for either you or the Committee on Equipment to realize how badly, nor to what extent, the present equipment in physics has been neglected, and allowed to run down. I have looked

Pressure of light experiment, Wilder Laboratory, 1903

over the collection carefully, and have found practically nothing of value which seems to have been acquired since Professor Young's time. There is much in the department which is interesting from an historical point of view, and some apparatus which after suitable repairs have been made may still be used, but I have never seen a department in worse condition.

Like Charles Young's, Nichols' initial wish list ($1,500) included primarily research apparatus, the precision thermometers, galvanometers, and dividing engine he would need to measure the pressure of light. He also was able to hire an assistant professor of physics, Gordon Ferrie Hull, who had graduated from the University of Toronto that had new teaching labs. He had earned his Ph.D. under A.A. Michelson at The University of Chicago. From 1899 until 1903, Nichols and Hull worked together

to confirm one of the more elusive predictions of James Clerk Maxwell's classic theory of electromagnetism (1873), viz., that light exerts pressure on objects, independent of heating effects. Resulting in important papers in leading American journals, this research embodied the high-precision techniques of turn-of-the-century experimental physics. It exploited the facilities of the new Wilder Laboratory and the latest European instruments that Nichols secured, drawing on relationships that he had established with German and Swiss makers during his years in Berlin. Like many of the first generation of American Ph.D. physicists, Nichols brought the practices of German physics to Dartmouth. Hull later wrote that "the years 1899-1903 were strenuous and exhilarating years for the department

233

of physics. Vacations were largely given up to research. In all of this work, Dr. Nichols was the invigorating spirit."[222]

In 1903, Nichols left Dartmouth to become chair of Columbia's physics department and director of its prestigious Phoenix Research Laboratory. In 1905-6, Hull spent a sabbatical year at the Cavendish Laboratory in Cambridge, England, studying the influence of electric fields on spectral lines. Upon his return to Dartmouth, he continued to equip the Wilder laboratories, took up teaching and essentially abandoned research. President Tucker's idea of the "historic college" would not nourish original investigations in experimental physics. Indeed, as they voted to accept Nichol's resignation, the trustees recorded "their appreciation of his important work in original research in physics and astrophysics—which has been carried on mainly outside the usual work for the College." Given such views and increasing enrollments, the College began finding other uses for the Wilder building, removing equipment from the special-purpose rooms and filling them with chalk-boards and desks. The Department of Mathematics and Astronomy later moved into the building, further restricting the space for physics. For the next four decades, Wilder Laboratory would serve primarily as a teaching facility. Even ambitious projects like the "Dartmouth accelerator" (see entry) of the late 1950s were designed for teaching rather than discovery. Not until 1965, when Dartmouth introduced small Ph.D. programs in the natural sciences, would research publications begin to trickle out of Wilder Laboratory.[224]

During Wilder Laboratory's first decade, Nichols and Hull spent about $11,500 for scientific equipment. Given Nichols' continental connections, much of the electrical and optical precision apparatus used for their research on the pressure of light came from German makers. Unlike Young in 1870-71, they purchased nothing from London. American makers primarily provided supplies for the student laboratories–stands, clamps, brackets, glassware, tubing, and tools.

They also purchased a 12-inch lathe and other equipment for the Wilder machine shop and hired a full-time mechanic. Hull reported in 1912: "We now have made in our laboratory instruments of precision which we previously had to purchase abroad. During the past two years there have been each year additions to the apparatus for the teaching of general physics. It has been necessary often to discard the apparatus of ten years ago for newer apparatus of better design. This constant attention to the teaching of general physics has given us an equipment well suited to that purpose."[225]

Much of the apparatus acquired over the next decades would be for the teaching of general physics, purchased commercially from American makers who were supplying the student laboratories then burgeoning at high schools, colleges, universities and medical schools across North America.

Table 3: Wilder Laboratory Instrument Purchases, 1898-1908[223]

Maker	Location	Purchases
C. Gerhardt	Bonn	$990
Keiser & Schmidt	Berlin	960
Schmidt & Haensch	Berlin	860
Wm Gaertner & Co.	Chicago	830
Siemens & Halske	Berlin	620
James G. Biddle	Philadelphia	540
A.J. Wilkinson & Co.		340
Eimer & Amand	New York	335
R. Fuess	Berlin-Steglitz	285
Otto Wolff	Berlin	255
Leeds & Northrup	Philadelphia	250
Société Genevoise	Geneva	200
Zeigler Electric Co.	Boston	180
Queen & Co.	Philadelphia	150
A.T. Thompson	Boston	135
W.G. Pye & Co.	Cambridge,	90

234

Firms like Leeds & Northrup, L.E. Knott, Eastern Science Supply Company, Weston Electrical Instrument Company, Central Scientific Company, The Scientific Shop, Welch Scientific and E.H. Sargent flourished during the first half of the twentieth century. At least ten percent of the items now in the King Collection derive from these makers.

Some precision apparatus would also be used for research, carrying on the tradition of careful physical measurement represented by the earlier pressure of light experiments. Between 1900 and 1964, over eighty M.A. theses were written in the Department of Physics, usually experimental, often exploring the behavior of simple instruments like a Geiger counter, frequency generator, X-ray tube, or photoelectric cell. Few of the theses resulted in scientific publication; yet more than a few of their authors entered careers in science, earning Ph.D. degrees elsewhere and assuming faculty positions at universities across America.

When Allen King arrived at Dartmouth in 1942, two years after Professor Hull had retired, he found Wilder Laboratory awash in these American instruments. Tucked away on back shelves and in attics across the campus, however, remained remnants of Dartmouth's earlier philosophical apparatus, layered in dust like archaeological strata. In these artifacts we find stories of science at an American college, of entrepreneurial professors, of competition with other institutions, of ambitious patrons, of commercial and local instrument making, of the rise and fall of research. Yes, the artifacts were used to study, measure and experiment; but over the years they assumed additional layers of significance in their local context.

While writing these lines on an August afternoon, we happen upon a scene at the loading dock behind Wilder Laboratory. Using large carts, two burly men wrestle cubic meters of old physics equipment from a storage area into a large dump truck. Removed from the room where apparatus goes when retired from active labs, these vacuum chambers, pipes, cables, racks of electronics equipment, and who knows what else are on their way to a salvage company, where they will be cut up for scrap copper, aluminum, iron and glass. Lacking storage space, we rescue for the King Collection a single glass flask, in which Airco Industrial Gases of Indianapolis shipped pure oxygen to the lab in the 1970s. Later that afternoon, another large truck arrives at the loading dock, bringing electronics equipment and a large vacuum chamber for an experiment in plasma physics now being assembled in Wilder Laboratory. On any given day, truckloads of scientific instruments enter and leave the campus. What gets saved and what has been saved for the historical record is a matter of space, personal preference, and luck.

Chapter 3

Engaging Instruments

In the first part of this book, we have sought to tell stories of individual instruments. We have examined makers and marketing, materials and design, provenance and acquisition, physical principles inherent in the objects, contexts of teaching and research, and changing patterns of their use. In this chapter, we want to draw back from the individual objects and take a wider view of scientific instruments in local university settings. We offer these musings not as dogmatic conclusions but rather as a stimulus to continued engagement with the material culture of science, with the objects used for study, measure and experiment on the natural world.

Taxonomies

Nineteenth-century instrument makers' catalogues and natural philosophy textbooks generally divide their contents into categories. Thus, we have chapters or sections for optics, magnetism, pneumatics, mechanics, etc. Professors organized their courses and lectures around these sections and they purchased apparatus and stored the apparatus using these general categories. Many university collections, such as ours, will probably have this implicit taxonomy imposed upon them, a scheme deriving simply from the manufacture and subsequent use of the instruments.

The first comprehensive inventory of Dartmouth's philosophical apparatus, drafted in 1862, groups the items into such broad, functional categories (see Table 1 Chapter 2) and presumably mirrors how the objects were arranged spatially in the storage areas. Likewise, the earliest photographic records of the apparatus, prepared for the 1876 exhibition, group the items into the same categories (see pages 129 and 163). One can find traces of the same taxonomy in the ordering of the hundred instruments in this book.

Clearly, these are pragmatic classifications, since from a practical standpoint instruments need to be named and ordered if they are to be organized and stored. However, not all instruments found in university collections can easily be placed into such pragmatic categories, especially as instruments from later in the twentieth century no longer fall neatly into the divisions of "classical" physics. We generally accept the name given a device by its manufacturer. Hence, our Strobotac (see entry), named by General Radio, is still known by that appellation, but what is its genus? Is it "optical," "electrical," or (where it is stored in our collection) "photographic?" Likewise, our Mann short-interval timer (see entry) could have been used in physics experiments on motion, the photography darkroom, or various chemical experiments. Allen King placed this instrument in a cabinet he labeled "time," a chapter not generally found in textbooks, but nonetheless a useful category enabling the storage of objects and the recovery of some information about their function. We do not know whether King agonized over such placements, or whether he asked if Strobotacs and interval timers might not be "instruments" at all but rather "tools" found in laboratories.

Historians of scientific instruments have proposed many taxonomies based on various markers. Social historians consider class backgrounds in distinguishing seventeenth-century mathematical and philosophical instruments. Philosophers interrogate the metaphysics of knowledge generated by the devices in positing essentialist categories such as observational versus experimental instruments or meters, graphs and scopes. Some collectors sort instruments by their constitutive materials, naming categories of brass, glass, ivory, wood or Bakelite. Historians of science might distinguish between instruments as models, pedagogical demonstrations and tools of research. Economic historians note differences between mass-produced, mass-marketed objects, one-off items produced by commercial makers for individual users, and unique pieces made by local craftsmen.[226]

At Dartmouth, we use the instruments of the King Collection as the basis for an undergraduate course. Each year our students are required to plan and prepare an exhibit. They must define a theme, select the items to display, write the labels, and decide how the individual instruments relate to each other. For their exhibits our students have used many taxonomies, often invented by themselves. One successful exhibit, "Is it Hot or Not?" categorized instruments by their "coolness" factor. "Coolness," as determined by a consensus of undergraduates (they established a mechanism by which their peers could vote), had little to do with the operation or design of an object but instead reflected the subjective, emotional response it evoked in the naïve viewer. "Coolness" worked as a coherent taxonomy; many other students attended the exhibit and found the nomenclature both comprehensible and persuasive.

University instrument collections, with traces of their previous, pragmatic taxonomies still intact, belie the notion of any single, essentialist, ontological set of categories into which scientific instruments "must" be placed. Chemical elements might have their periodic table, living organisms their biological taxonomy. But the historical significance of scientific instruments, made, used and collected for many purposes, can better be explored if we eschew their immutable placement into rigid categories.

Compound instruments

Beyond taxonomy, university collections as cultural attics also focus attention on how separate instruments interact together when employed in the classroom or research laboratory. Many of the instruments featured in this book are "stand alone" apparatus, not designed to be integrated into ensembles that we call compound instruments. A sundial, microscope, galvanometer, circumferentor or polarimeter generally have been used as independent objects for purposes of studying, measuring or experimenting. Some instruments, however, contain subunits that, according to convention, could be classified as instruments in their

The King Collection in use: Physics 7, 2003

238

own right. For example, the ESSCO spectra demonstrator (see entry) contains Bunsen-style burners, a spectroscope and viewing telescope. As a compound instrument, the demonstrator requires the inclusion of these separate instruments for its operation.

Other instruments in this book require an assemblage of other instruments before they can begin to "work." The electric egg or the Coolidge X-ray tube (see entries) produce interesting phenomena only when connected to a battery and an induction coil (see entries). Likewise, the Franklin bells require a Leyden jar, charged by an electric machine (see entries), to demonstrate their effects. An air pump, condensing chamber and manometer (see entries) also constitute a compound instrument. The laser needs its power supply (see entry). In each of these cases, an individual instrument would stand mute without its cast of related instruments. Only as compound instruments do the components interact to yield the desired results.

The Shattuck Observatory can be similarly viewed. At its most basic core, we have a telescope and building and dome to house it. The building must protect the telescope from the weather, yet also permit access to the sky. We might not ordinarily call an observatory building an instrument; yet it is an operational part of the telescope not unlike the mount. Associated too are transit telescopes, clock drives, chronometers, chronographs, spectroscopes, photographic apparatus and the many bits and pieces needed to allow the telescope to "stretch from some object to an observer." Some of these components are capable of performing "stand alone" work as independent instruments; but when associated with the rest of the instruments in an observatory, they become organs in a scientific organism.

In its initial organization, the Shattuck Observatory included a library of books with the glass and brass of its telescopes and mechanical components. Today the Shattuck houses a set of the first Palomar sky survey plates, conducted from 1950-57 on the 48-inch Schmidt telescope, and some earlier glass plates from Harvard surveys. The function of the observatory in generating knowledge depends upon such collections of images and texts, and they too are organs in the observatory as a compound instrument.

Extending the scale even further, we might consider high-energy particle accelerators, kilometers in diameter, as another example of compound instruments. Many of the components of these massive ensembles are instruments in their own right; other components lose their "instrumentness" when removed from the accelerator.

Compound instruments, not a category discernable in the organization of Professor King's cabinets, nonetheless provide a useful means of understanding complex associations within and between some instruments of science.

Making instruments

By the 1770s, when Dartmouth started acquiring scientific apparatus, instrument making, especially in Britain, had become located within a set of specialized trades. Over the previous century, skilled craftsmen who had once worked as founders, armourers, silversmiths, gunsmiths, locksmiths, clockmakers, dialists, furniture makers, or opticians began to serve the growing commercial market for the apparatus of natural philosophy and the mathematical instruments of surveying. As historian Anita McConnell has shown, increased demand resulted in the production of more-or-less standard models of instruments.[227] A customer seeking a microscope, for example, could purchase "off-the-shelf" a Culpeper model rather than an individually designed object. Although individual Culpepers might differ slightly, they were nonetheless recognizable types or models. Among the American instrument makers, introduction of standard models did not occur until early in the nineteenth century.

Many of the objects in the King Collection, and indeed many of the items featured in this book, were purchased as off-the-shelf instruments. The early British apparatus, such as the orrery, Gilbert's mechanical powers, Jones' achromatic telescope, and Margetts' chronometer (see entries), were surely models, made in large numbers for late-eighteenth-century buyers. Likewise, items by early American makers—Kennard's surveying quadrant, Pool's circumferentor, Young's Wye level, James Green's manometer and Chamberlain's air pump (see entries)—appear to have been widely available. As the number of American colleges and universities expanded rapidly after 1800, their demand for philosophical apparatus stimulated American makers to increase their offering of standard models. By the 1840s American makers frequently supplied standard packages of instruments, especially selected to outfit classrooms for given topics such as chemistry or mechanics.[228]

Yet university contexts, where professors can experiment with off-the-shelf demonstration apparatus or modify instruments during research investigations, also can stimulate the design and construction of innovative items that are not commercially available. Such nonstandard objects can be specially ordered from commercial makers or can be built in local shops. The King Collection also features such instruments. The Salem maker, John Prince, probably cobbled together Dartmouth's electric machine (see entry) from parts scavenged from various objects he imported from Britain. Prince did not have a shop producing standard models of such a machine. Likewise, the electrotome (see entry) that we have

cautiously traced to Charles Page's workshop, may have been a roughly assembled prototype for a device that Page and Daniel Davis, Jr. later would offer in a standard version. The double-pass, multi-prism spectroscope (see entry) arose through collaboration between Alvan Clark and Professor Charles Young, and would remain a custom-built rather than off-the-shelf item in the Clarks' offering.[229] Likewise, Louis Bell's seismograph (see entry) was custom-made. Although carefully constructed by a skilled machinist, Bell's instrument probably was made for use in his consulting business and was not distributed commercially.

Other university instruments have a local provenance. The Hanover machinist and "general jobber" John N. Brown signed two objects now in the King Collection, an optical square and a tangent galvanometer (see entries). Our galvanometer is marked "No. 2," which may suggest that Brown built at least several versions of the device. However, nothing in the local business directories suggests that Brown regularly made scientific instruments. Presumably, he built the items in response to special requests by Dartmouth faculty. Professor King's elastometer and the Mann short-interval timer (see entries) were constructed in Dartmouth shops by machinists employed by the College; both are unique, local objects that would not enter commercial production. The Dartmouth accelerator was another unique construction, although the only remaining component of that compound instrument is a commercially made diffusion pump (see entry).

Regardless of whether produced in multiple copies or as unique objects, instruments come into being, as do all products of human artifice, through teleological practices. Their makers bring together selected materials, designs, tools, skill, aesthetics, work, and assumptions about economics, value and future users to create devices that serve various purposes. Most of these processes and the numerous decisions they entail become virtually invisible once a completed object leaves the workshop or factory. Given the proprietary nature of manufacturing processes, instrument makers have often been reticent to publicize the particular secrets of their trade. Their published catalogues describe finished products, not the tools, techniques and steps required to make the products. Even patent documents can conceal more than they reveal. Nonetheless, close examination of the finished artifacts may provide clues about how they were made.

Like many artifacts, scientific instruments often display features such as richly finished wooden surfaces, embellished components, handsomely designed structural elements, where simpler shapes and lesser surfaces would suffice. As David Pye, master wood craftsmen and longtime professor of furniture design at the Royal College of Art, London, has noted, proponents of utilitarian analysis refer to such features as "useless work," as effort invested that does not enhance the successful function of the artifact. Pye argues, however, that "useless work" expresses a fundamental component of craftsmanship.[230]

Varying degrees of craftsmanship are seen in the objects presented in this book. The variable resistance coil (see entry) has been assembled with a minimum of "useless work." The instrument performs its electical function, but not much effort has been expended to enhance its aesthetic appeal. Professor King's elastometer (see entry), on the other hand, is a nicely finished piece of locally built apparatus. Its smooth, neatly painted surfaces required extra work, although they do not necessarily contribute to the ability of this device to demonstrate elastic phenomena. Louis Bell's seismograph (see entry), another custom made instrument, illustrates even more attention to visual detail. Most of the parts are turned from solid brass bars and plates. Then the machinist decorated the flat top surface using a hand-held scraper, work that does not augment the operation of the instrument.

Designers, too, introduce "useless work." The student volt-ammeter by L.E. Knott (see entry) proudly displays its Art Nouveau embellishments that contribute nothing to the measuring function of the instrument. Items as mundane as Bunsen burners were also designed to enhance their attractiveness, as the Central Scientific Company emphasized in its catalogue (see entry). Margett's chronometer (see entry) contains obvious examples of "useless work" in its design. Even within the normally unseen interior, its supporting spindles are handsomely turned and polished.

Elegantly detailed work might well help manufacturers sell instruments in competitive markets or serve additional functions implicit in the artifacts. Scientific instruments, both past and present, are replete with "useless work," regardless of whether they were made for eighteenth-century drawing rooms where they served a social function or twentieth-century university laboratories where well-crafted instruments still have aesthetic appeal. Thus, "useless work" is not useless. It reflects the many functions that instruments play in the different worlds they inhabit. Humans embellish what they make.

References

DCLSC = Dartmouth College Library, Special Collections

1 Ernie Jorgenson, *Date of Manufacturing by Serial Number, Marchant Rotary Calculators, Monroe Rotary Calculators, Comptometer Key Drive Calculators* (Lewiston, Idaho: Office Machines of America, 2002) unpag.

2 *Typewriter Topics* June 1929: 48.

3 Charles A. Holden, "The Uses of Calculating Machines," *Engineering News* 45 (1901): 405.

4 Edwin Thacher, *Thacher's Calculating Instrument or Cylindrical Slide-Rule, Containing Complete and Simple Rules and Directions for Performing the Greatest Variety of Useful Calculations with Unexampled Rapidity and Accuracy* (New York: D. Van Nostrand, 1884) 3-4; Edwin Thacher, *Directions for Using Thacher's Calculating Instrument* (New York: Keuffel & Esser Co., 1907) 3-4; Bob Otnes, "Thacher Notes," *Journal of the Oughtred Society* 2 (1993): 21-25; Wayne E. Feely, "Thacher Cylindrical Slide Rules," *Chronicle of the Early American Industry Association* 50 (1997): 125-27.

5 Central Scientific Company, *Scientific Instruments, Laboratory Apparatus and Supplies for Physics, Chemistry, Biological Sciences, and Industrial Testing, Catalogue J136* (Chicago: Neely Printing Company, 1936) 84-87.

6 Lyman Spalding, *A New Nomenclature of Chemistry, Proposed by Messrs de Morveau, Lavoisier, Berthollet and Fourcroy* (Hanover, New Hampshire: Moses Davis, 1799); Denis I. Duveen and Herbert S. Klickstein, "The Introduction of Lavoisier's Chemical Nomenclature into America," *Isis* 45 (1954): 371-72; I. Bernard Cohen, *Some Early Tools of American Science* (New York: Russell & Russell, 1950) 76; Herbert Darling Foster, "Webster and Choate in College: Dartmouth under the Curriculum of 1796-1819," *Dartmouth Alumni Magazine* 19 (1927): 512.

7 Cyrus Perkins, "Chemical Apparatus Now in the Laboratory Belonging to Dartmouth College," 1815, ms., DCLSC, MS 815528.4.

8 King Collection Archive.

9 Allen L. King, et al., "Elasticity of Soft Body Tissues," *Scientific Monthly* 71 (1950): 258-60; Allen L. King, *Thermophysics* (San Francisco: W. H. Freeman and Co., 1962) 133; Allen L. King, *My Life: The Making of a Teacher-Scholar* (Concord, New Hampshire: Common Sense Press, 2003) 138-41.

10 Quoted by Henry A. Rowland, letter to Daniel C. Gilman, undated [summer 1875], D.C. Gilman Papers Ms. 1, Milton S. Eisenhower Library, Special Collections, The Johns Hopkins University.

11 Franklin B. Dexter, ed., *Extracts from the Itineraries and Other Miscellanies of Ezra Stiles, D.D., Ll.D., 1755-1794* (New Haven: Yale University Press, 1916) 396; John K. Lord, ed., *A History of Dartmouth College, 1815-1909* (Concord, New Hampshire: Rumford Press, 1913) 573-74.

12 "Course of Studies at Dartmouth College," *Dartmouth Gazette* 27 September 1805: 4; George Adams, *Lectures on Natural and Experimental Philosophy*, 4 vols. (Philadelphia: Whitehall, 1807) 3: 263-65: Alan Q. Morton and Jane A. Wess, *Public & Private Science: The King George III Collection* (Oxford: University Press, 1993) 432-33.

13 "Memorandum of Statement of Louis Bell," 1912, ts., King Collection Archive.

14 Edward F. Connelly, *Machine Tool Reconditioning and Applications of Hand Scraping* (St. Paul: Machine Tool Publications, 1955) 168.

15 Silvio Bedini, *Early American Scientific Instruments and Their Makers* (Washington, D.C.: Museum of History and Technology, 1964) 70-72, 109-11, 39; Charles E. Smart, *The Makers of Surveying Instruments in America since 1700* (Troy, New York: Regal Art Press, 1962-67) 64-65. For surveying practices in eighteenth-century New England, see Jerald E. Brown, *The Years of the Life of Samuel Lane, 1718-1806: A New Hampshire Man and His World* (Hanover, New Hampshire: University Press of New England, 2000).

16 Sanborn C. Brown and Leonard M. Rieser, *Natural Philosophy at Dartmouth: From Surveyor's Chains to the Pressure of Light* (Hanover, New Hampshire: University Press of New England, 1974) 6-12; Foster, "Webster and Choate," 512; *Catalogue of the Officers and Students of Dartmouth College for the Academical Year 1873-4* (Hanover, New Hampshire: no publ., 1873) 37.

17 Hillier sale 1993, reproduced in *Maine Antique Digest* May 1993: 24a; Skinner sale 1997, reproduced in *Historical New Hampshire* 52 (1997): 83; Currier Gallery of Art, accession 1959.4.15, reproduced in Smart, *Surveying Instruments* 231.

18 James Hill Fitts, *History of Newfields, New Hampshire, 1638-1911* (Concord: Rumford Press, 1912) 337, 569; Smart, *Surveying Instruments* 92, 191, 230-31; Bedini, *Early American Scientific Instruments* 124-29; Gerald D. Foss, *Three Centuries of Freemasonry in New Hampshire* (Somersworth, New Hampshire: New Hampshire Publishing Co., 1972) 393; James L. Garvin and Donna-Belle Garvin, *Instruments of Change: New Hampshire Hand Tools and Their Makers* (Canaan, New Hampshire: Phoenix Publishers, 1984).

19 Jennifer L. Anderson, "Nature's Currency: The Atlantic Mahogany Trade and the Commodification of Nature in the Eighteenth Century," *Early American Studies* 2 (2004): 47-80.

20 *Whitely's Philadelphia Annual Advertiser* for 1820, quoted in National Museum of American History,

Physical Sciences Collection, Surveying and Geodesy, 20 August 2004, <http://americanhistory2.si.edu/surveying/object.cfm?recordnumber=742296 >; see also Deborah Jean Warner, "The Surveyor's Compass," *Rittenhouse* 1 (1987): 64-76.

21 Donald Wing and Anne Wing, "The Pool Family ofEaston, Massachusetts," *Rittenhouse* 4 (1990): 118.

22 *Rittenhouse* 1 (1987): 90; Bedini, *Early American Scientific Instruments* 97-104; Robert Vogel and Edmund Hands, "The Pools of Easton, Massachusetts," *Chronicle of the Early American Industry Association* 50 (1997): 1-11.

23 *Adjustments of Gurley Levels* (Troy, New York: W. & L.E. Gurley, 1927); W. &. L.E. Gurley, *A Manual of the Principal Instruments Used in American Engineering and Surveying Manufactured by W. & L.E. Gurley [1874 Ed.]*, Reprint ed. (Mendham, New Jersey: Astragal Press, 1993) 119-30.

24 Deborah Jean Warner, "William J. Young: From Craft to Industry in a Skilled Trade," *Pennsylvania History* 52 (1985): 53-68; Robert C. Miller, "Dating Young Instruments," *Rittenhouse* 5 (1990): 21-24.

25 Edward C. Lathem, ed., *Jeremy Belknap: Journey to Dartmouth in 1774* (Hanover, New Hampshire: Dartmouth Publications, 1950); Grace S. Machemer, "Headquartered at Piscataqua: Samuel Holland's Coastal and Inland Surveys, 1770-74," *Historical New Hampshire* 57 (2002): 19; Brown and Rieser, *Natural Philosophy at Dartmouth* 18; Leon Burr Richardson, *History of Dartmouth College* (Hanover: Dartmouth College Publications, 1932) 121-22.

26 "A Catalogue of Mathematical, Philosophical and Optical Instruments as made and sold by Thomas Heath and Tycho Wing" [1771], bound with *The Practical Surveyor,* eds. John Hammond and Samuel Warner, 4th ed. (London: T. Heath & Wing, 1765) 15: Robert C. Strong, "Samuel Holland, Esquire, Servant of the Crown," *Dartmouth Alumni Magazine* 21.February (1929): 241-42; Allen L. King, "The Case of the Missing Sundial," *Dartmouth College Library Bulletin* 8 (1968): 66-69; Allen L. King, "In Quest of a Gnomon," *Dartmouth College Library Bulletin* 14 (1973): 18-23; Allen L. King, "End of a Quest," *Dartmouth College Library Bulletin* 15 (1975): 69-70; King, *My Life* 185-86.

27 Peter I. Drinkwater, *The Art of Sundial Construction,* 4th ed. (Shipston-on-Stour: P. Drinkwater, 1996) 66; Frederick Chase, *A History of Dartmouth College and the Town of Hanover, New Hampshire. Edited by John K. Lord* (Cambridge: John Wilson and Son, University Press, 1891) 289; Lord, ed., *Dartmouth College* 607; Samuel Johnson, *A Dictionary of the English Language,* 4th ed., vol. 2 (Dublin: Thomas Ewing, 1775). The use of modern terms to refer to old alloys is discussed by Justine Bayley, "Alloy Nomenclature," *Medieval Finds from Excavations in London, 3: Dress Accessories, c. 1150-c. 1450,* eds. Geoff Egan and Frances Pritchard (Woodbridge: Boydell Press, 2002) 13-17.

28 Jonathan Betts, email to David Pantalony, 14 January 2004; Brooks Palmer, *The Book of American Clocks* (New York: Macmillan Publishing Co., 1950) 279; Daybook, 1833-35, ms. notebook, William Bond & Son and the Bond Family, 1724-1931, Scientific Instrument Collection, Harvard University, box 4, folder 2. We thank Jonathan Betts for drawing our attention to this latter source.

29 Peter Ifland, *Taking the Stars* (Malabar: Krieger Publishing Company, 1998) 172-73.

30 Richard H. Goddard, letter to Air Technical Service Command, 7 March 1947, King Collection Archive; Papers of the United States Naval Training School at Dartmouth, 1942-45, DCLSC, DA-11; Ray Nash, *Navy at Dartmouth* (Hanover, New Hampshire: Dartmouth Publications, 1946); James G. Schneider, *The Navy V-12 Program: Leadership for a Lifetime* (Boston: Houghton Mifflin, 1987).

31 Robert N. Buck, *North Star over My Shoulder: A Flying Life* (New York: Simon & Schuster, 2002) 386-87.

32 Fred Springer-Miller, letter to NN, 14 July 1994. King Collection Archive.

33 Lord, ed., *Dartmouth College* 253, 571-74; Brown and Rieser, *Natural Philosophy at Dartmouth* 19-23;Trustees of Dartmouth College, letter to Ira Young, 20 August 1834, Treasurer's Papers, DCLSC, DA-2(4):11.

34 John R. Millburn, *Adams of Fleet Street: Instrument Makers to King George III* (Aldershot: Ashgate, 2000); Anita McConnell, "From Craft Workshop to Big Business: The London Scientific Trade's Response to Increasing Demand, 1750-1820," *London Journal* 19 (1994): 36-53.

35 Alice Walters, "Importing Science in the Early Republic: Union College's 'First Purchase' of Instruments and Books," *Rittenhouse* 16 (2002): 101.

36 John R. Millburn, "Nomenclature of Astronomical Models," *Bulletin of the Scientific Instrument Society* 34 (1992): 7-9; Dexter, ed., *Extracts* 396.

37 Eastern Science Supply Company, *ESSCO Rapid Reference Pictorial Bulletin, RR1* (Boston: Eastern Science Supply Company, 1932).

38 Richard H. Goddard, letter to S. L. Boothroyd, 12 June 1940, King Collection Archive.

39 Undated newspaper clippings and advertisements, C.J. Kullmer Papers, Archives and Records Management, Syracuse University; "Kullmer Constellation Finder [Advertisement]," *The Public* 13 (1910): 1199; "Equatorial Star Finder," *Scientific American* 103 (1910): 281; C.J. Kullmer, *Star Maps and Star Facts*, 3d improved ed. (Syracuse: C.J. Kullmer, 1911).

244

40 William Jones, *The Description and Use of a New Portable Orrery ... A Catalogue of Optical, Mathematical and Philosophical Instruments Made and Sold by William and Samuel Jones*, 4th ed. (London: W. and S. Jones, 1794) 2.

41 William Simms, *The Achromatic Telescope and its Various Mountings* (London: Troughton & Simms, 1852) 26; Walters, "Importing Science," 85-107.

42 Elias Loomis, *Recent Progress of Astronomy Especially in the United States*, 3d ed. (New York: Harper & Brothers, 1856) 292; W. C. Rufus, "Astronomical Observatories in the United States Prior to 1848," *Scientific Monthly* 19 (1924): 120-39; David F. Musto, "A Survey of the American Observatory Movement, 1800-1850," *Vistas in Astronomy* 9 (1967): 87-92; Derek Howse, "The Greenwich List of Observatories," *Journal for the History of Astronomy* 17.4 (1986): 84-89; Craig H. White, "Natural Law and National Science: The 'Star of Empire' in Manifest Destiny and the American Observatory Movement," *Prospects* 20 (1995): 119-60.

43 Trustee Minutes, July 1841, 28 July 1846, DCLSC, DA-1, vol. 3: 7, 64-65; Faculty Committee, letter to trustees, 27 July 1846, DCLSC, MS 846427.1; Ira Young, letters to trustees, 28 July 1846, 27 July 1847, 23 July 1849, 23 July 1850, DCLSC, Shattuck Observatory Papers, DA-9(1):12; *Catalogue of the Officers and Students of Dartmouth College, 1849-50*, (Troy, New York: J. C. Kneeland's Steam Press, 1849) 27; Allen King notes, July 1969, Collection Archive. Young's comparison neglected the 9.5-inch Merz refractor at the U.S. Naval Observatory since 1842.

44 Ira Young, letters to George C. Shattuck, 2 October 1852, 5 October 1852; John Aiken, letter to trustees, 12 December 1852, Trustee Minutes, 23 December 1852, DCLSC, DA-1, vol. 3: 152-70; Ira Young, letter to Oliver P. Hubbard, 12 May 1853; [Ira and Charles Young], "Description of a Part of the Observatory of Mr. Lassell of Liverpool, England," undated, unsigned ms.; misc. receipts and correspondence regarding construction of Shattuck Observatory, 1847-53, DCLSC, Shattuck Observatory Papers, DA-9(1):4; [Ammi Young], undated, unsigned architectural drawings, DCLSC, D.C. History, Iconography 225; Charles A. Young, "Diary of a Trip to Europe," 1853, autograph ms., Charles A. Young Papers, DCLSC, ML-49(1):14, 10-11, 58, 74, 81, 91, 141-42, 45, 48; William Lassell, "Description of an Observatory Erected at Starfield, near Liverpool," *Memoirs of the Royal Astronomical Society* 12 (1842): 265-72; Leon Burr Richardson, "Shattuck Observatory: A Biography of the College's Oldest Science Building," *Dartmouth Alumni Magazine* December 1943: 15-16, 72; Brown and Rieser, *Natural Philosophy at Dartmouth* 44-56; Deborah Jean Warner and Robert B. Ariail, *Alvan Clark & Sons: Artists in Optics*, 2d ed. (Richmond, Virginia: Willmann-Bell, Inc., 1995) 79-82; David Pantalony, "The Purchasing Trip of Ira and Charles Young in 1853," *Bulletin of the Scientific Instrument Society* 76 (2003): 23-27.

45 Ira Young, "Report on the Purchase of Books and Apparatus in Europe," undated ms., DCLSC, MS 852940; Ira Young, letter to trustees, 28 July 1846, DCLSC, Shattuck Observatory Papers, DA-9(1):12; Allen L. King, "The Search," *Dartmouth College Library Bulletin* 12 (1971): 18-22.

46 Myles W. Jackson, *Spectrum of Belief: Joseph von Fraunhofer and the Craft of Precision Optics* (Cambridge: MIT Press, 2000) 172-74; "The New Spectroscope and its Results," *The Dartmouth* 4 (1870): 353; "Alvan Clark and His Telescopes," *Boston Journal of Chemistry* 6 (1871): 16; Charles A. Young, "The Color Correction of Certain Achromatic Object Glasses," *American Journal of Science* 19 (1880): 454; Warner and Ariail, *Alvan Clark & Sons* 82; "The New Telescope," *The Dartmouth* 5 (1871): 324-25; J. E. Nourse, "Observatories in the United States, II," *Harper's New Monthly Magazine* 49 (1874): 523-25; Henry Fairbanks, letter to Asa Smith, 25 July 1871, DCLSC, MS 871425; Richard H. Goddard, letter to Warner and Swasey Company, 26 March 1938, King Collection Archive; King, "The Search," 18-22.

47 John Merrill Poor, "Notes on Changes Made at Shattuck Observatory During the Latter Part of Dr. Tucker's Administration as President," undated ts., Shattuck Observatory Papers, DCLSC, DA-9(1):7; "On the Hill: Shattuck Observatory," *Dartmouth Alumni Magazine* December 1946: 19-21; Edwin B. Frost, "Observations of Comet A1890, Made at the Shattuck Observatory of Dartmouth College with the 9.4-Inch Equatorial and Ring-Micrometer," *Astronomical Journal* 10 (1890): 32; "Shattuck Observatory Shown Acquiring a New Spring Chapeau," *Dartmouth Alumni Magazine* June 1959: 30; Allyn J. Thompson, "Come Michaelmas," *Sky & Telescope* 14 (1955): 259; George Z. Dimitroff and James G. Baker, *Telescopes and Accessories* (Philadelphia: Blakiston Company, 1945) 220; James R. Willard, letter to Walter Barrows, 26 January 1865, DCLSC, MS 865126; Edwin B. Frost, *An Astronomer's Life* (Boston: Houghton Mifflin Company, 1933) 90.

48 Henry J. Green, *Scientific Instruments* (Brooklyn: no publ., [after 1949]) 17, 22.

49 *Dartmouth Alumni Magazine* April 1947: 47.

50 Poor, "Shattuck Observatory," 2-3.

51 C. A. Robert Lundin, "Account Books," 1896-1928, ms. notebooks, C. A. Robert Lundin Sr. and Jr. Papers, Center for American History, University of Texas at Austin, Box 3F87, 126. We are indebted to John W. Briggs for drawing our attention to this source.

52 J. A. Bennett, *Science at the Great Exhibition* (Cambridge: Whipple Museum of the History of Science, 1983) 3-5; George P. Bond and R. F. Bond, "Description of an Apparatus for Making Astronomical Observations by Means of Electro-Magnetism," *Report of the British Association for the Advancement of Science, Sections* (1851): 21-22.

53 *The Dartmouth* 6 (1872): 217; Charles A. Young, "On a Proposed Printing Chronograph," *American Journal of Science* 42 (1866): 99-104.

54 S. W. Stratton, *Measurement of Time and Tests of Timepieces*, Circular of the National Bureau of Standards, No. 51 (Washington, D.C.: Government Printing Office, 1914) 21.

55 "Apparatus of the Appleton Professorship of Natural Philosophy," undated ms., DCLSC, MS 000737, 11; Charles A. Young, "Spectroscopic Observations," 1870-72, ms. notebook, DCLSC, ML-49(1):19, 66.

56 Josiah P. Cooke, Jr., "An Improved Spectroscope," *Chemical News* 8 (1863): 8; Joseph P. Cooke, Jr., "On the Construction of a Spectroscope with a Number of Prisms," *American Journal of Science* 40 (1865): 305-13; Warner and Ariail, *Alvan Clark & Sons* 79-81; J. A. Bennett, "The Spectroscope's First Decade," *Bulletin of the Scientific Instrument Society* 4 (1984): 3-6; Klaus Hentschel, *Mapping the Spectrum: Techniques of Visual Representation in Research and Teaching* (Oxford: Oxford University Press, 2002) 89-98; Alvan Clark, letter to C.A. Young, 15 March 1866, Young Papers, DCLSC, ML-49(6):59; Charles A. Young, "Pocket Diary," 1866, ms. notebook, DCLSC, ML-49(1):8, entries for January and October; Trustee Minutes, 16 July 1866, DCLSC, DA-1, vol. 3: 401 (approved $350); Dartmouth College Treasurer, "Appleton Fund Accounts," 1862-75, ledger, Treasurer's Papers, DCLSC, DA-2(81):12, unpag. (expended $450). The accessories now associated with this spectroscope (page 81) were assembled by Allen King in 1963. See Allen L. King, "Centennial for a Spectrometer," *Dartmouth Alumni Magazine* March 1966: 34; Charles A. Young, "Spectroscopic Notes," *Journal of the Franklin Institute* 58 (1869): 141-42; Charles A. Young, "On a New Method of Observing Contacts at the Sun's Limb," *American Journal of Science* 48 (1869): 370-78; Charles A. Young, "Spectrum Observations at Burlington, Iowa, During the Eclipse of August 7, 1869," *Proceedings of the American Association for the Advancement of Science* 18 (1870): 78-84; Karl Hufbauer, *Exploring the Sun: Solar Science since Galileo* (Baltimore: Johns Hopkins University Press, 1991) 62, 112-14. During the 1869 eclipse, William Harkness of the U.S. Naval Observatory also saw one green coronal line, which he measured at 1497 K and also attributed to iron vapor. See William Harkness, "Appendix 2, Reports of the Observations of the Total Eclipse of the Sun, August 7, 1869," *Astronomical and Meteorological Observations Made at the U.S. Naval Observatory During the Year 1867* (1870): 60-67.

57 Charles A. Young, "Photograph of a Solar Prominence," *American Journal of Science* 50 (1870): 404-5; Charles A. Young, "Report of My Observations of the Eclipse of December 22, 1870," *U.S. Coast and Geodetic Survey Report for 1870, Appendix No. 16*, ed. Benjamin Peirce (Washington, D.C.: Government Printing Office, 1871) 30-31; Charles A. Young, "Spectroscopic Notes," *Journal of the Franklin Institute* 60 (1870): 332. Young's brief "A New Form of Spectroscope," dated 3 October 1870, appeared in the *Journal of the Franklin Institute* 60 (1870), 331-40; *Chemical News* 22 (1870), 277-80; and *Nature* 3 (1870), 110-13. See "Spectroscopic Research," *Boston Journal of Chemistry* 5 (1871): 92; Brown and Rieser, *Natural Philosophy at Dartmouth* 68-83; A. J. Meadows, *Early Solar Physics* (Oxford: Pergamon Press, 1970) 221-22; Alex Soojung-Kim Pang, *Empire and the Sun: Victorian Solar Eclipse Expeditions* (Stanford: Stanford University Press, 2002).

58 Otto von Littrow, "Ueber eine neue Einrichtung des Spectralapparates," *Sitzungsberichte der kgl. Akademie der Wissenschaften, math.-naturwiss. Classe* 47 (1863): 26-32; Lewis M. Rutherfurd, "On the Construction of the Spectroscope," *American Journal of Science* 39 (1865): 129-32; John Browning, "On a Spectroscope in which the Prisms are Automatically Adjusted for the Minimum Angle of Deviation for the Particular Ray under Examination," *Monthly Notices of the Royal Astronomical Society* 30 (1870): 198-202; Young, "Spectroscopic Notes," 334; Heinrich Schellen, *Spectrum Analysis in its Application to Terrestrial Substances and the Physical Constitution of the Heavenly Bodies [1871]*, trans. Jane Lassell and Caroline Lassell, ed. William Huggins (London: Longmans, Green & Co., 1872) 417-43; J. Normal Lockyer, *Contributions to Solar Physics* (London: Macmilland and Co., 1874) 578-85.

59 Charles A. Young, "Spectroscopic Notes, on the Construction, Arrangement and Best Proportions of the Instrument with Reference to its Efficiency," *Journal of the Franklin Institute* 62 (1871): 351-53; John Browning, "On an Universal Automatic Spectroscope," *Monthly Notices of the Royal Astronomical Society* 32 (1872): 213-14; John Browning, *How to Work with the Spectroscope* (London: John Browning, 1878) 12-17, 50-51; Nicolaus von Konkoly, *Handbuch für Spectroscopiker im Cabinet und am Fernrohr* (Halle: Wilhelm Knapp, 1890) 221-28, 333-44; Warner and Ariail, *Alvan Clark & Sons* 208; Charles A. Young, "Note on the Use of a Diffraction 'Grating' as a Substitute for the Train of Prisms in a Solar Spectroscope," *American Journal of Science* 5 (1873): 472-73. Priority for inventing the two-pass solar spectroscope design has been contested. In his initial published description of October 1870, Young allowed that "after planning the instrument I learned that the same idea of sending the light twice through the prisms … had also occurred to Mr. Lockyer and others," at which point the editor of *Nature* (Lockyer himself!) tartly added a note: "An instrument exactly similar in all essentials to the one here described has been used by Mr. Lockyer for more than a year past." However in an 1874 summary of his work, Lockyer, the leading British astrophysicist, illustrated only Young's two-pass spectroscope, not his own, adding in a footnote: "I had a spectroscope with a prism of this kind constructed in 1869, and afterwards the same idea occurred independently to Professor Young." Never did Lockyer publish any spectroscopic observations taken with a two-pass instrument. In 1870, the Dublin maker Howard Grubb sent William Huggins, another leading British spectroscopist, an "unfinished" two-pass device consisting of five compound prisms (see direct-vision spectroscopes entry, page 87). Grubb claimed that this arrangement produced a total dispersion of about 90 degrees, "probably the largest ever obtained" (roughly 25 percent greater than the dispersion of Young's two-pass device). Yet Huggins also did not publish any observations made with a two-pass spectroscope. And the earliest two-pass design, Otto von Littrow's 1863 instrument (apparently unknown to Young and Lockyer), which employed a mirror to reflect the beam back on itself, also yielded no published results. See Charles A. Young, "Spectroscopic Notes," *Nature* 3 (1870): 110; Charles A. Young, "Spectroscopic Notes, on the Construction, Arrangement, and Best Proportions of the Instrument with Reference to its Efficiency," *Nature* 5 (1871): 86; Howard Grubb, "Automatic Spectroscope for Dr. Huggins' Sun Observations," *Monthly Notices of the Royal Astronomical Society* 31 (1870): 36-38; Lockyer, *Solar Physics* 167-68; Littrow, "Ueber eine neue Einrichtung des Spectralapparates," 26-32. For useful descriptions of the panoply of late-nineteenth-century prism spectroscopes, see J. Scheiner, *Die Spectralanalyse der Gestirne* (Leipzig: Wilhelm Engelmann, 1890) 64-120; Konkoly, *Handbuch* 169-351; Heinrich Kayser, *Handbuch der Spectroscopie*, 8 vols. (Leipzig: Hirzel, 1900-34) 1: 489-547.

60 Young, "Spectroscopic Observations," 66-67.

61 Schellen, *Spectrum Analysis* 113-20; Gerard L'Estrange Turner, *The Great Age of the Microscope* (Bristol: Adam Hilger, 1989) 328-29; Deborah Jean Warner, "Direct Vision Spectroscopes," *Rittenhouse* 7 (1992): 40-48.

62 Browning, *Spectroscope 1878.*

63 Eastern Science Supply Company, *Rapid Reference, 1932.*

64 Gordon Ferrie Hull, "The Investigation of the Influence of Electrical Fields upon Spectral Lines," *Astrophyical Journal* 25 (1907): 14.

65 George Sweetnam, *The Command of Light: Rowland's School of Physics and the Spectrum* (Philadelphia: American Philosophical Society, 2000); Deborah Jean Warner, "Rowland's Gratings," *Vistas in Astronomy* 29 (1986): 125-30.

66 Louis Bell, "The Absolute Wave-Length of Light," *American Journal of Science* 35 (1888): 275.

67 J.S. Ames, letters to Gordon Ferrie Hull, 1 and 5 December 1899, Hull Papers, DCLSC, ML-47(5):14.

68 *Illustrated Price List of Physical and Mechanical Instruments Made by the Société Genevoise* (Geneva: W. Kündig & Son, 1900) 10-11.

69 Gerard L'Estrange Turner, *Nineteenth-Century Scientific Instruments* (Berkeley: University of California Press, 1983) 299-300; "Appleton Apparatus," 16.

70 *Physiological Instruments Manufactured by the Cambridge Scientific Instrument Company, Ltd.*, (Cambridge: University Press, 1899) 114-15; Hermann von Helmholtz, *Handbuch der physiologischen Optik*, 2d revised ed. (Leipzig: L. Voss, 1896) 667-68; Casey Wood, ed., *American Encyclopedia and Dictionary of Ophthalmology*, 18 vols. (Chicago: Cleveland Press, 1913-21) 6860; Lucien Howe, *Muscles of the Eye*, 2 vols. (New York: G. P. Putnam's Sons, 1907-8) 1: 180-86.

71 W.H. Walmsley & Co., *A Classified and Illustrated Price-List of Photographic Cameras, Lenses, and Other Apparatus and Materials for the Use of Amateur and Professional Photographers* (Philadelphia: W.H. Walmsley, 1884); R. & J. Beck, *Revised Illustrated Catalogue of Microscopes, Object-Glasses and Apparatus, Microtomes, Mounting-Materials and All Requisites for Microscopical Work, Manufactured by R. And J. Beck ... London, Sole American Agents W.H. Walmsley and Co.* (Philadelphia: no publ., 1885); *Gopsill's Philadelphia City Directory* (Philadelphia: James Gopsill's Sons, 1878-93); Andrew H. Eskind, ed., *International Photography: George Eastman House Index to Photographers, Collections, and Exhibitions*, 3d enlarged ed., 3 vols. (New York: G.K. Hall, 1998) s.v. "Walmsley"; W.H. Walmsley, "A Handy Photomicrographic Camera," *Proceedings of the American Society of Microscopists* 12 (1890): 69-74.

72 Moses Dyer Carbee and his cousin, Samuel P. Carbee, practiced medicine together from 1874 to 1882. The latter had received his M.D. from Dartmouth Medical School in 1866. King Collection Archive; Henry Morton and Coleman Sellers, "Memoir of Joseph Zentmayer," *Journal of the Franklin Institute* 126 (1888): 488; Turner, *Microscope* 161-63.

73 Maison Jules Duboscq, *Historique & Catalogue de tous les instruments d'optique supérieure* (Le Mans: Edmond Monnoyer, 1885) 3; "Appleton Apparatus," 14; Adolphe Ganot, *Traité élémentaire de* *physique expérimentale et appliquée et de météorologie,* 6th ed. (Paris: L'auteur-éditeur, 1856) 451-52; Allan A. Mills, "Portable Heliostats (Solar Illuminators)," *Annals of Science* 43 (1986): 372.

74 Paolo Brenni, "Soleil, Duboscq and Their Successors," *Bulletin of the Scientific Instrument Society* 51 (1996): 7-16; Pantalony, "Purchasing Trip," 23-27; Young, "Diary of a Trip to Europe," 45.

75 Schmidt & Haensch invoice, 28 November 1871, Appleton Professorship Papers, DCLSC, DA-173, folder 3; Dartmouth College Treasurer, "Appleton Fund Accounts," 37; *Illustrated Scientific and Descriptive Catalogue of Achromatic Microscopes Manufactured by J. & W. Grunow & Co.* (New Haven: T.J. Stafford, 1857) 85-93.

76 Dietrich Tutzke, "Georg Grübler und Karl Holborn," *Wissenschaftliche Zeitschrift der Karl-Marx-Universität Leipzig, math.-naturwiss. Reihe* 21 (1972): 341-48; Michael Titford, "Comparison of Historic Grübler Dyes with Modern Counterparts," *Biotechnic & Histochemistry* 76 (2001): 23-30.

77 Pantalony, "Purchasing Trip," 23-27.

78 Albert G. Ingalls, ed., *Amateur Telescope Making,* 2d ed. (New York: Scientific American Publishing Co., 1928); Russell W. Porter, "The Telescope Makers of Springfield, Vermont," *Popular Astronomy* 31 (1923): 153-62; Berton C. Willard, *Russell W. Porter, Arctic Explorer, Artist, Telescope Maker* (Freeport, Maine: Bond Wheelright Co., 1976).

79 *Boston Directory* (Boston: Sampson & Murdock Company, 1895-1920); Bruno Kolbe, "Ein handlicher Lichtbrechungs-Apparat," *Zeitschrift für den physikalischen und chemischen Unterricht* 9 (1896): 20-24; Hans Hartl, "Neue physikalische Apparate," *Zeitschrift für den physikalischen und chemischen Unterricht* 9 (1896): 113-16; Max Kohl A.G., *Physical Apparatus, Price List No. 50* (Chemnitz: 1911) 488-93; L.E. Knott Apparatus Company, *Catalogue of Scientific Instruments, Catalogue 17* (Boston: L.E. Knott Apparatus Company, 1912) 321-22; L.E. Knott Apparatus Company, *Catalogue of Scientific Instruments, Catalogue 21* (Boston: L.E. Knott Apparatus Company, 1916) 252-53; L.E. Knott Apparatus Company, *Harcourt Equipment for Physics: Standard for Education since 1895* (Cambridge: L.E. Knott Apparatus Co., 1928); Central Scientific Company, *Catalogue M, Physical and Chemical Apparatus for Science Laboratories* (Chicago: no publ., 1909) 224; Central Scientific Company, *Catalogue J136* 1461-63; "Mystery Object: Hartl's Disk," *Bulletin of the Scientific Instrument Society* 55 (1997): 29.

80 Dayton Clarence Miller, *Laboratory Physics: A Student Manual for Colleges and Scientific Schools* (Boston: Ginn and Company, 1903) 255-61; Richard Staley, "Travelling Light," *Instruments, Travel and Science: Itineraries of Precision from the Seventeenth*

to the Twentieth Century, eds. Marie-Noëlle Bourguet, Christian Licoppe and H. Otto Sibum (London: Routledge, 2002) 243-72.

81 Francis W. Sears and Mark W. Zemansky, *University Physics,* 3d ed. (Reading, Massachusetts: Addison-Wesley, 1964) 906; Francis W. Sears, Mark W. Zemansky and Hugh D. Young, *University Physics,* 7th ed. (Reading, Massachusetts: Addison-Wesley, 1987) 927.

82 W. & L.E. Gurley, *Physical and Scientific Instruments* (Troy, New York: W. & L.E. Gurley, 1910) 144; Trustee Minutes, 24 May 1907, DCLSC, DA-1, vol. 5: 430.

83 John Lee Comstock, *A System of Natural Philosophy,* 218th ed. (New York: Pratt, Oakley & Co., 1860) 146-47; Deborah Jean Warner, "Air Pumps in American Education," *The Physics Teacher* 25 (1987): 82-85.

84 *Catalogue of Pneumatic Instruments Manufactured and Sold by N.B. & D. Chamberlain* (Boston: Benjamin Perkins, 1844) iv, 6-7; *Ritchie's Illustrated Catalogue of Philosophical Instruments and School Apparatus* (Boston: Andrew Holland, 1860) 16; L.E. Knott Apparatus Company, *Catalogue of Scientific Instruments, Catalogue 26* (Boston: L.E. Knott Apparatus Company, 1921) 74; Deborah Jean Warner, "Compasses and Coils: The Instrument Business of Edward S. Ritchie," *Rittenhouse* 9 (1994): 2.

85 *Catalogue of Pneumatic Instruments, 1844,* 26-27.

86 *The Dartmouth* 3 (1869) 399; "Appleton Apparatus," 3.

87 *Ritchie's Catalogue of Physical Instruments* (Boston: E.S. Ritchie & Sons, 1878) 26.

88 *Ritchie's Catalogue,* unpag. preface.

89 Greg Drake, "Nineteenth-Century Photography in the Upper Connecticut Valley: An Annotated Checklist," *Dartmouth College Library Bulletin* 25 (1985): 76; United States Centennial Commission, *International Exhibition 1876,* 9 vols. (Washington, D.C.: Government Printing Office, 1880-84) 8: 59.

90 Ernest Fox Nichols, "The Wilder Physical Laboratory of Dartmouth College," *Physical Review* 12 (1901): 366-71.

91 Central Scientific Company, *Catalogue J136* 1575; Deborah Jean Warner, email to Kremer, 11 April 2005.

92 *Hanover Gazette* 14 July 1955; Frank Pemberton, "Dartmouth Physicists Bombard Deuterium," *Industrial Science and Engineering* November 1955: 6-9; Frank Pemberton, "New Hampshire's First Nuclear Accelerator," *New Hampshire Profiles* November 1955: 22-24; Robert J. Grainger, "The Yield and Energy Distribution of Neutrons from the Dartmouth Accelerator," MA thesis, Dartmouth College, 1956; Harold C. Britt, "The Absorption and Detection of

Fast Neutrons," MA thesis, Dartmouth College, 1958.

93 David Pantalony, "Seeing a Voice: Rudolph Koenig's Instruments for Studying Vowel Sounds," *American Journal of Psychology* 117 (2004): 425-42; David Pantalony, "Analyzing Sound in the Nineteenth Century: The Koenig Sound Analyzer," *Bulletin of the Scientific Instrument Society* 68 (2001): 16-21; Paolo Brenni, "The Triumph of Experimental Acoustics: Albert Marloye (1795-1874) and Rudolph Koenig (1832-1901)," *Bulletin of the Scientific Instrument Society* 44 (1995): 13-17.

94 *The New Grove Dictionary of Music and Musicians*, 2d ed. (New York: Grove, 2001), s.v. "organ"; Hermann Helmholtz, *Die Lehre von den Tonempfindungen als physiologische Grundlage für die Theorie der Musik* (Braunschweig: F. Vieweg und Sohn, 1863) 154; Julien Fau, *Nouveau manuel complet du physicien-préparateur*, 2 vols. (Paris: Roret, 1853) 1: 404; Hermann von Helmholtz, *On the Sensations of Tone as a Physiological Basis for a Theory of Music*, trans. Alexander J. Ellis, 2d English ed. (New York: Longmans, Green, 1885) 95-96.

95 Albert Marloye, *Catalogue des principaux appareils d'acoustique* (Paris: Bonaventure et Ducessois, 1851) 43.

96 Ganot, *Physique 1856* 202.

97 Hans Christian Ørsted, "On Acoustic Figures [1807]," *Selected Scientific Works of Hans Christian Ørsted*, eds. Karen Jelved, Andrew D. Jackson and Ole Knudsen (Princeton: Princeton University Press, 1998) 261; Félix Savart, "Recherches sur les vibrations normales," *Annales de chimie et de physique* 36 (1827): 187-208; Michael Faraday, "On a Peculiar Class of Acoustical Figures," *Philosophical Transactions* 121 (1831): 314-35; Marloye, *Catalogue 1851* 46; Fau, *Nouveau manuel* 397-400; Ganot, *Physique 1856* 180-81.

98 Oliver Wendell Holmes, "My Hunt after 'The Captain' [1862]," *Works*, vol. 8 (Boston: Houghton, Mifflin and Company, 1892) 19.

99 Young, "Diary of a Trip to Europe," 54. For more on Marloye, see Brenni, "Experimental Acoustics," 13-17.

100 André Propser Crova, "Description d'un appareil pour la projection mécanique des mouvements vibratoires," *Annales de chimie et de physique, 4th sér.* 12 (1867): 288-308; Jules Lissajous, "Instruments d'astronomie, de géodésie, de topographie, de marine, d'optique et d'acoustique," *Exposition Universelle de Paris 1867, Rapports du jury international*, ed. Michel Chevalier, vol. 2 (Paris: Paul Dupont, 1868) 480-84; Julian Holland, "Charles Wheatstone and the Representation of Waves," *Rittenhouse* 13, 14 (2000): 86-106, 27-46; Max Kohl A.G., *Price List No. 50* 416, iii; Ernest Fox Nichols, "Instrument Purchases," 1898-1908, ms. notebook, King Collection Archive.

101 Willem Hackmann, *Electricity from Glass: The History of the Frictional Electrical Machine, 1600-1850* (Alphen aan den Rijn: Sijthoff & Noordhoff, 1978) 124-38; Deborah Jean Warner, "Edward Nairne: Scientist and Instrument-Maker," *Rittenhouse* 12 (1998): 72-77.

102 Dexter, ed., *Extracts* 396; Charles W. Upham, "Memoir of Rev. John Prince," *American Journal of Science* 31 (1837): 211; Adams' receipt, February 1814, Treasurer's Papers, DCLSC, DA-2(1):73; Trustee Minutes, August 1816, DCLSC, DA-1, vol. 2: 96.

103 Charles A. Young, letter to John Lord, 26 February 1889, quoted in Lord, ed., *Dartmouth College* 608 ("… a considerable quality of the old apparatus came from the sale (or gift) of Dr. Prince of Salem, who I think was a friend of Priestley's. The cylindrical electrical machine, and the large Franklin 36 jar battery of Leyden jars were always mentioned as having belonged to the Prince collection."); Henry Fairbanks, "Appleton Fund Inventory," 1862, autograph ms., Treasurer's Papers, DCLSC, DA-2(81):12, 19 ("36 1-pt jars, covered box, Dr. Priestley's used by Franklin"); "Appleton Apparatus," 15-16 ("Battery of 36 jars - each 1/2 sq ft … formerly belonged to Priestley, was used by Franklin for about two years, was purchased by Dr. Prince of Salem, Ms. At the sale of Priestley's apparatus & on Dr. Prince's death was acquired by the College, at the same time with the cylinder machine & a good deal of other apparatus"); Allen L. King, "Two Curious Footnotes," *Dartmouth College Library Bulletin* 7 (1966): 13-14; Allen L. King, "A Case of Mistaken Identity," *Dartmouth College Library Bulletin* 9 (1968): 29-32; Sara Schechner, "John Prince and Early American Scientific Instrument Making," *Publications of the Colonial Society of Massachusetts* 59 (1982): 502-3; Ronald K. Smeltzer, "The Library and Apparatus of John Prince," *Rittenhouse* 1 (1986): 97.

104 Central Scientific Company, *Catalogue J136* 1273.

105 King, "Mistaken Identity," 29-32. Henry Fairbanks was the Appleton Professor of Natural Philosophy at Dartmouth from 1859 to 1865.

106 Frank L. Pope, *Modern Practice of the Electric Telegraph*, 11th ed. (New York: Russell Brothers, 1869) 15-19.

107 Gustav Wiedemann, *Die Lehre von der Elektricität*, 3 vols. (Braunschweig: Vieweg und Sohn, 1882-83) 1: 729-95; Adolphe Ganot, *Elementary Treatise on Physics, Experimental and Applied for the Use of Colleges and Schools*, trans. E. Atkinson, 15th ed. (New York: William Wood and Company, 1901) 820-22.

108 Daniel Davis, Jr., *Descriptive Catalogue of Apparatus and Experiments* (Boston: Marden & Kimball, 1838) 45-49; Charles G. Page, "Magneto-

249

Electric and Electro-Magnetic Apparatus and Experiments," *American Journal of Science* 35 (1839): 258-59; Daniel Davis, Jr., *Manual of Magnetism* (Boston: Daniel Davis, Jr., 1842) 190-91; Daniel Davis, Jr., *Manual of Magnetism*, 3d ed. (Boston: Daniel Davis, Jr., 1851) 304-6.

109 Niels H. de V. Heathcote, "N.J. Callan: Inventor of the Induction Coil," *Annals of Science* 21 (1965): 145-67; Willem Hackmann, "The Induction Coil in Medicine and Physics, 1835-1877," *Studies in the History of Scientific Instruments*, eds. Christine Blondel, Françoise Parot, Anthony Turner and Mari Williams (London: Rogers Turner Books, 1989) 235-50. For somewhat polemical reviews of the early history of the induction coil, see Charles G. Page, *The American Claim to the Induction Coil and its Electrostatic Developments* (Washington, D.C.: Intelligencer Printing House, 1867); J. A. Fleming, *The Alternate Current Transformer in Theory and Practice*, 2 vols. (London: *The Electrician* Printing and Publishing Company, 1889-92) 2: 1-35. Page's "original instrument …, made for him by Daniel Davis, of Boston, in 1838" (Page, *American Claim* 93) is at the National Museum of American History (cat. 309,254). We have examined other Page-Davis coils in Harvard's Collection of Historic Scientific Instruments, Allegheny College, and the personal collection of John Jenkins (two examples).

110 Heathcote, "N.J. Callan," 145-67; Charles G. Page, "Method of Increasing Shocks," *American Journal of Science* 31 (1837): 137-41; Charles G. Page, "Researches in Magnetic Electricity," *American Journal of Science* 34 (1838): 364-73; Charles G. Page, "On the Use of the Dynamic Multiplier," *American Journal of Science* 32 (1837): 354-60; Charles G. Page, "New Magnetic Electrical Machine of Great Power," *American Journal of Science* 34 (1838): 163-69; Charles G. Page, "Magnetic Electrepeter and Electrotome," *American Journal of Science* 35 (1839): 112-13; Page, *American Claim* 61, 92.

111 Fairbanks, "Appleton Fund Inventory"; "Appleton Apparatus," 19.

112 Fleming, *Transformer* 2: 29-70; Hackmann, "Induction Coil," 244-50; Paolo Brenni, "Heinrich Daniel Ruhmkorff (1803-1877)," *Bulletin of the Scientific Instrument Society* 41 (1994): 4-8; E.S. Ritchie, "On a Modified Form of Ruhmkorff's Induction Apparatus," *American Journal of Science* 24 (1857): 45-46; *Ritchie's Catalogue*, 102; "The Induction Coil," *English Mechanic and Mirror of Science* 4 (1867): 413-14, 28-29.

113 Charles A. Young, "Induction Coil of Unusual Size," *Journal of the Franklin Institute* 60 (1870): 7.

114 Apps invoice, 2 February 1871, Appleton Professorship Papers, DCLSC, DA-173(3); Young, "Spectroscopic Observations," 34, 66-67; *The Great Lightning Inductorium Exhibited at the Polytechnic Institution ... Manufactured by A. Apps* (London: Alfred Apps, 1869); William Spottiswoode, "Description of a Large Induction-Coil," *Philosophical Magazine, ser. 5* 3 (1877): 30-34; "Appleton Apparatus," 19.

115 Allen L. King, "The Misleading Nameplate," *Dartmouth College Library Bulletin* 10 (1969): 15-18. For details of how Apps wound his secondary coils and improved the interruptor, see Fleming, *Transformer* 2: 59-61.

116 *Preisverzeichnis von Dr. H. Geissler Nachfolger Franz Müller*, 9th ed. (Bonn: no publ., 1904) 405, 88-89.

117 Arthur-Auguste de La Rive, *Treatise on Electricity in Theory and Practice*, trans. Charles. V. Walker, 3 vols. (London: Longman, Brown, Green, Longmans & Roberts, 1853-58) 2: 307-10, 3: 292-95; Adolphe Ganot, Atkinson, 2d ed. (London: Longmans, Green, and Co., 1867) 604, 721-26; Apps invoice, 2 February 1871, Appleton Professorship Papers, DCLSC, DA-173(3).

118 Adolphe Ganot, *Elementary Treatise on Physics, Experimental and Applied for the Use of Colleges and Schools*, 13th ed. (London: Longmans, Green, and Co., 1890) 931.

119 Allen & Hanburys, *Catalogue of Surgical Instruments and Appliances, Aseptic Hospital Furniture, Ward Requisites, Etc.* (London: Allen & Hanburys, 1901) 827.

120 Dr. Stöhrer & Sohn, *Preis-Verzeichniss von Apparaten und Instrumenten für den physikalischen Unterricht an mittleren und höheren Lehranstalten* (Leipzig: O. Leiner, 1892) 142-43; Edwin B. Frost, "Experiments on the X-Rays," *Science* N.S. 3 (1896): 235-36; Edwin B. Frost, "The First X-Ray Experiment in America?," *Dartmouth Alumni Magazine* 22 (1930): 383-84; Peter K. Spiegel, "The First Clinical X-Ray Made in America–100 Years," *American Journal of Roentgenology* 164 (1995): 241-43; James Klaas, "The Early History of X-Rays," *Dartmouth Undergraduate Journal of Science* 5 (2002): 42-44.

121 W. D. Coolidge, "A Powerful Roentgen Ray Tube with a Pure Electron Discharge," *Physical Review, ser. 2* 2 (1913): 409-30; Lewis Gregory Cole, "Technique and Experimental Application of Hard Rays for Deep Röntgentherapy," *Surgery, Gynecology and Obstetrics* 21 (1915): 522-56; Eddy Clifford Jerman, *Modern X-Ray Technic* (St. Paul: Bruce Publishing Company, 1928); "Coolidge Tube 50 Years Old," *Radiologic Technology* 39 (1968): 296-97; Robert G. Arns, "The High-Vacuum X-Ray Tube," *Technology and Culture* 38 (1997): 852-90.

122 Lloyd C. Douglas, *Doctor Hudson's Secret Journal* (Boston: Houghton Mifflin Company, 1939) 89.

123 C.T.W. Wilson, "On a Sensitive Gold-Leaf Electrometer," *Proceedings of the Cambridge Philosophical Society* 12 (1904): 135-39.

124 Adolphe Ganot, *Elementary Treatise on Physics, Experimental and Applied for the Use of Colleges and*

Schools, trans. E. Atkinson, 17th ed. (London: Longmans, Green, and Co., 1906) 836-37.

125 Hamilton Child, *Gazetteer of Grafton County, N.H., 1709-1886* (Syracuse, New York: Syracuse Journal Company, Printers, 1886) 300; *Catalogue of Dartmouth College, 1889-90* (Concord, New Hampshire: Republican Press Association, 1889) 46.

126 Robert L. Barclay, *The Art of the Trumpet-Maker* (Oxford: Oxford University Press, 1992) 107; Hans E. Wulff, *The Traditional Crafts of Persia* (Cambridge: MIT Press, 1966) 24.

127 Paolo Brenni, "Louis François Breguet and Antonie Breguet," *Bulletin of the Scientific Instrument Society* 50 (1996): 19-24; Breguet, *Catalogue illustré: Appareils et matériaux pour la télégraphie électrique, instruments divers, électricité - physique mécanique - météorologie, physiologie*, 3d ed. (Paris: Typographie Lahure, 1877) 49.

128 Thomson did not describe his astatic mirror galvanometer in print until William Thomson, "Galvanometers for the Measurement of Electrical Currents and Potentials [1884]," *Mathematical and Physical Papers* (Cambridge: University Press, 1882-1911) 2: 442-48; H. R. Kempe, "The Thomson Galvanometer," *Telegraphic Journal and Electrical Review* 2 (1874): 241-44; Siemens & Halske, *Wissenschaftliche Messinstrumente* (Berlin: Springer, 1881) A-7; Henri DuBois and Heinrich Rubens, "Modificirtes astatisches Galvanometer," *Annalen der Physik* N.F. 48 (1893): 236-51; Max Kohl A.G., *Price List No. 50* 882-83. George Green and John T. Lloyd, *Kelvin's Instruments and the Kelvin Museum* (Glasgow: University of Glasgow, 1970) must be used with caution.

129 Ernest Fox Nichols and Gordon Ferrie Hull, "The Pressure Due to Radiation," *Physical Review* 17 (1903): 50.

130 John Burnett, "The Origins of the Electrocardiograph as a Clinical Instrument," *The Emergence of Modern Cardiology*, eds. W. F. Bynum, C. Lawrence and V. Nutton (London: Wellcome Institute for the History of Medicine, 1985) 53-76; M. J. G. Cattermole and A. F. Wolfe, *Horace Darwin's Shop: A History of the Cambridge Scientific Instrument Company* (Bristol: Hilger, 1987) 74-75, 108, 220-30; Robert F. Frank, Jr., "The Telltale Heart: Physiological Instruments, Graphic Methods, and Clinical Hopes, 1854-1914," *The Investigative Enterprise*, eds. William Coleman and Frederic L. Holmes (Berkeley: University of California Press, 1988) 250-63. For an illustration of a slightly earlier Hindle string galvanometer, mounted as part of a complete electrocardiograph unit now at the Bakken Museum (acc. 80.1.43), see Albert W. Kuhfeld, "The Graphic Method," *Rittenhouse* 9 (1995): 96.

131 Leeds & Northrup Company, *Apparatus for Capacitance, Inductance and Magnetic Measurements* (Philadelphia: Leeds and Northrup Company, 1927) 12. We thank Thomas Greenslade for providing a copy of this trade catalogue.

132 Alexander G. Medlicott, Jr. and John E. Kent, eds., *Class of 1950, Dartmouth College 50 Year Book* (Sheenboro, Quebec: David L. Prentice, 2000) 339-40.

133 Nichols, "Instrument Purchases," January 1900.

134 Graeme Gooday, "The Morals of Energy Metering," *The Values of Precision*, ed. M. Norton Wise (Princeton: Princeton University Press, 1995) 239-82.

135 Quoted in Diane Chalmers Johnson, *American Art Nouveau* (New York: H. N. Abrams, 1979) 14.

136 L.E. Knott Apparatus Company, *Catalogue 21*; L.E. Knott Apparatus Company, *Catalogue 26*; Richard Guy Wilson, Dianne H. Pilgrim and Dickran Tashjian, *The Machine Age in America, 1918-1941* (New York: Brooklyn Museum, 1986).

137 *Students' Potentiometer, Bulletin No. 765* (Philadelphia: Leeds & Northrup Company, 1926) 3.

138 Cambosco Scientific Company, *Cambosco Order Book, 1950-51* (Boston: Cambosco, 1950) back cover.

139 John W. Howell and Henry Schroeder, *History of the Incandescent Lamp* (Schenectady: The Maqua Co., 1927).

140 Dartmouth Scientific Association, Minute book, 1883-1909, ms., DCLSC, DO-6(1); Charles Franklin Emerson, "The Seven Founders: An Address Commemorating the Fiftieth Anniversary of the Dartmouth Scientific Association," 1920, ts., DCLSC, DC Hist Q11 .D3.E42.

141 William T. Doyle, email to Kremer, 15 September 2004; Sears and Zemansky, *University Physics 1964* 79, 135, 43, 274, 302-3, 497; "New Eyes for Modern Industry," *General Radio Experimenter* September 1960: 3-11; Robert Hunt, *Elementary Physics* (London: Reeve and Benham, 1851) 232.

142 Friedrich Kurylo, *Ferdinand Braun: A Life of the Nobel Prizewinner and Inventor of the Cathode-Ray Oscilloscope*, trans. Charles Susskind (Cambridge: MIT Press, 1981).

143 V.J. Phillips, *Waveforms: A History of Early Oscillography* (Bristol: Hilger, 1987) 238-49.

144 Theodore Soller, Merle A. Starr and George E. Valley, Jr., eds., *Cathode Ray Tube Displays* (New York: McGraw-Hill Book Co., 1948) 10. For an early, once-classified description, see F.J. Gaffney, *Instruction Manual: Browning Type 'a' Synchronizer* (Cambridge: Radiation Laboratory, MIT, 1941).

145 King, *My Life* 175. Catalogue descriptions for King's optics course first mention "coherence" as a topic in 1961-62.

146 Andreas Urs Sommer, Dagmar Winter and Miguel Skirl, *Die Hortung: Eine Philosophie des Sammelns* (Düsseldorf: Parerga, 2000); Brian Neville and Johanne Villeneuve, eds., *Waste-Site Stories: The Recycling of Memory* (Albany: State University of

New York Press, 2002); Werner Muensterberger, *Collecting, an Unruly Passion* (Princeton: Princeton University Press, 1994).

147 Alfred F. Whiting, inventory and letter to Allen King, 2 July 1962; Newell Grant, letter to W. James King, 9 May 1963, both in King Files, King Collection Archive. According to accession records now in the Hood Museum of Art, Dean William Kimball of the Thayer School and William Doyle, an assistant professor of physics, initiated these transfers.

148 John Stewart, letter to Allen King, 28 January 1964, King Files, King Collection Archive; *The Dartmouth* 2 April 1964: 1, 4; Allen King, "Scientific Revolution Show, 1964," binder of photographs and labels, King Collection Archive; Allen L. King, *My Life: The Making of a Teacher-Scholar* (Concord, New Hampshire: Common Sense Press, 2003) 180. Although the show initially was announced as "scientific instruments dating from the Age of Shakespeare," King in fact had no artifacts from before 1770 and thus extended the chronological sweep of the exhibit, following A. Rupert Hall, *The Scientific Revolution, 1500-1800* (London: Longmans, Green, 1954).

149 Allen L. King, "Two Curious Footnotes," *Dartmouth College Library Bulletin* 7 (1966): 13-14. For King's "Summer Project on Old Apparatus, Report," 30 September 1967 and other related documents, see King Files, King Collection Archive.

150 Press release, King Vertical File, DCLSC; Allen L. King, *Dartmouth College: From Natural Philosophy to Contemporary Science, 1769-1969* (Hanover, New Hampshire: Dartmouth College, 1969); *Dartmouth College Bicentennnial Exhibit: Historical Philosophical Apparatus, a Pictorial Review,* (Hanover, New Hampshire: Dartmouth College, 1970).

151 Deborah Jean Warner, "Some Collections of Scientific Instruments in the U.S.," *Rittenhouse* 7 (1993): 97-105. See Thomas S. Greenslade's many reports of "visits to collections" in *Rittenhouse.*

152 S. Blume, "What Ever Happened to the String and Sealing Wax?," *Invisible Connections*, eds. R. Bud and S. Cozzens (Bellingham: SPIE, 1992) 87-101; M. Lourenço, "Musées et collections des universités," *Musée des arts et métiers, La revue.* 41 (2004): 51-60.

153 Hood Museum of Art, Dartmouth College, accession 772.1.30192; Lord, ed., *Dartmouth College* 601-2; W. Wedgwood Bowen, *A Pioneer Museum in the Wilderness* (Hanover, New Hampshire: Dartmouth College Museum, 1958) 1-17; "A List of the Curiosities to Be Found in the Museum at D. College [Copy after 1920]," c. 1810, Museum Vertical File, DCLSC.

154 "Catalogue of Books Belonging to Dartmouth College Library," 1775, ts., Woodward Room Vertical File, DCLSC; Lois A. Krieger, *The Woodward Succession: A Brief History of the Dartmouth College Library, 1769-2002* (Hanover, New Hampshire: Dartmouth College Library, 2002).

155 *Boston Evening Post*, 9 September 1771, 7 October 1771, 21 October 1771; *Boston News-Letter,* 24 October 1771. The term Philosyllogôn does not appear in classical Greek literature, as indexed by the *Thesaurus linguae graecae.* According to Frederick Chase, *A History of Dartmouth College and the Town of Hanover, New Hampshire, Edited by John K. Lord* (Cambridge: John Wilson and Son, University Press, 1891) 237, this anonymous friend of the College was Aaron Hutchinson (Yale 1747), a polemical Massachusetts clergyman known for his proficiency in Greek and Latin and who in 1774 would move to Vermont. In 1780, he received an honorary degree from Dartmouth. See Rush C. Hawkins, *A Biographical Sketch of the Rev. Aaron Hutchinson* (New York: Art Age Press, 1888).

156 Trustee Minutes, May 1771, DCLSC, DA-1, vol. 1: 21; Eleazar Wheelock, letter to Phillips, 30 September 1774, DCLSC, Ms 774530.2; Eleazar Wheelock, *A Continuation of the Narrative of the Indian Charity School ... Now Incorporated with Dartmouth College* (S.l.: No publ., 1772) 19-20; Chase, *Dartmouth College passim.* In 1769, Governor Wentworth had visited Harvard's library and philosophical apparatus, "where some experiments were done for his entertainment." Quoted in Vera M. Butler, "Education as Revealed by New England Newspapers Prior to 1850," Ph.D., Temple University, 1935, 21. For Wentworth's assessment of the 1771 episode, see Chase, *Dartmouth College* 274.

157 Wentworth, letter to Wheelock, 28 October 1774, transcribed in "Selections from Portfolios in Various Libraries," *Historical Magazine, and Notes and Queries Concerning the Antiquities* 5 (1869): 388. The letter names the maker "Ramsey." Yet Gloria Clifton, *Directory of British Scientific Instrument Makers, 1550-1851* (London: Zwemmer, 1995) includes no British maker by this name and it seems highly likely that Wentworth was referring to Jesse Ramsden. Trustee Minutes, May 1773, August 1779, DCLSC, DA-1, vol. 1: 24-25, 45.

158 Trustee Minutes, September 1783, DCLSC, DA-1, vol. 1: 77; John Wheelock, "An Account of My Tour in Europe Relative to the Prosperity of Dartmouth College," 1784, DCLSC, Ms 784157; Chase, *Dartmouth College* 573-74; Leon Burr Richardson, *History of Dartmouth College* (Hanover: Dartmouth College Publications, 1932) 208; Jane Marcou, *Life of Jeremy Belknap, D.D., the Historian of New Hampshire* (New York: Harper and Brothers, 1847) 71. Invented in 1731 by John Hadley, "Hadley's quadrant" eventually became known as the navigational sextant or octant. An octant by Spencer, Browning & Rust remains in the King Collection.

159 In 1773, Dartmouth awarded Hurd an honorary degree. Three years later, Hurd broke with Dartmouth by refusing to join the "successionist movement" instituted by College people to shift Hanover and several neighboring towns from New Hampshire to Vermont. By late 1782, that movement collapsed and Bezaleel Woodward, a Dartmouth faculty member who had led the rebellion, was appointed justice of the peace by the New Hampshire state government. One might speculate that the government of New Hampshire, with whom Hurd had preserved close ties, might have used Hurd and his quadrant as a symbolic means to thank the College for returning to the fold. See Jere R. Daniell, *Experiment in Republicanism: New Hampshire Politics and the American Revolution, 1741-1794* (Cambridge: Harvard University Press, 1970) 145-61.

160 We are unable to identify, with certainty, an eighteenth-century optical instrument maker by this name. Perhaps "Baker's microscope" refers to a solar microscope, known to be at Dartmouth since 1774 and fully described and nicely illustrated in Henry Baker, *The Microscope Made Easy,* 4th ed. (London: R. and J. Dodsley, 1754) 22-26. This book, in Dartmouth's initial library, may have provided a local name for the solar microscope.

161 Franklin B. Dexter, ed., *Extracts from the Itineraries and Other Miscellanies of Ezra Stiles, D.D., Ll.D., 1755-1794* (New Haven: Yale University Press, 1916) 396. Eighteenth-century sources had long recognized both Hadley and the Philadelphia glazier, Thomas Godfrey, as inventors of the double-reflecting "quadrant," later known as the sextant. See James Logan, "An Account of Mr. T. Godfrey's Improvement of Davis's Quadrant," *Philosophical Transactions* 38 (1734): 441-50.

162 In 1808, Deerfield Academy, located about 80 miles south of Dartmouth in northwestern Massachusetts, announced the purchase from W. & S. Jones of London of an air pump, sextant, theodolite, 2 1/2-foot achromatic telescope, microscope, electric machine, pair of globes and planetarium, a list that mirrors the Dartmouth collection at that time. See John McKnight, "Philosophical Apparatus at William and Mary," *Rittenhouse* 3 (1989): 64-69; E. Robert Paul, "Electrostatic Apparatus in the Nineteenth-Century American College," *Rittenhouse* 2 (1988): 105-8; Sara Schechner, "Tools for Teaching and Research: John Prince, the Deerfield Academy, and Educational Reform in the Early Republic," *Rittenhouse* 10 (1996): 98-100; V. Ennis Pilcher, *Early Science and the First Century of Physics at Union College, 1795-1895* (Schenectady: Coneco Litho Graphics, 1994); Alice Walters, "Importing Science in the Early Republic: Union College's 'First Purchase' of Instruments and Books," *Rittenhouse* 16 (2002): 85-107; I. Bernard Cohen, *Some Early Tools of American Science* (New York: Russell & Russell, 1950).

163 Trustee Minutes, September 1782, DCLSC, DA-1, vol. 1: 73; N. Lord and W. Wheeler, draft report, undated [circa 1828], Faculty Minutes, DCLSC, DA-175.1(1), folder "Faculty Laws, 1821-1901." See Jer-

emy Belknap, *The History of New Hampshire,* 3 vols. (Boston: Belknap and Young, 1784-92) 3: 296; "Course of Studies at Dartmouth College," *Dartmouth Gazette* 27 September 1805: 4; Herbert Darling Foster, "Webster and Choate in College: Dartmouth under the Curriculum of 1796-1819," *Dartmouth Alumni Magazine* 19 (1927): 509-16, 605-16.

164 Trustee Minutes, August 1790, August 1801, DCLSC, DA-1, vol. 1: 136-37, 252; Chase, *Dartmouth College* 584. In 1787, Yale's president had described a 30-inch achromat; the instrument currently in the King Collection is 42 inches in length. Both telescopes are offered in W. & S. Jones' catalogue of 1794. Presumably the College had first acquired the shorter achromat, and then sometime after 1787 purchased the longer version.

165 The national survey of magnetic deviation was not published until the 1830s. See Ira Young, letter to Elias Loomis, 11 April 1839, Loomis Papers, Yale University, Beinecke Rare Book and Manuscript Library, Z117.00113 v. 12; Elias Loomis, "On the Variation and Dip of the Magnetic Needle in Different Parts of the United States," *American Journal of Science* 34 (1838): 293, 301, 306.

166 George Adams, *Lectures on Natural and Experimental Philosophy,* 4 vols. (Philadelphia: Whitehall, 1807) 4: unpag. end matter. It is not clear who raised the subscriptions for this massive edition. See Paul Carle, "Le cabinet de physique et l'enseignement des sciences au Canada français: Le cas du Seminarie de Québec et de l'Université Laval (1633-1920)," Ph.D., Université de Montréal, 1986, 73.

167 Adino N. Brackett, letter to his parents, 5 September 1841, and Sullivan C. Kimball, surveying notebook, September 1857, DCLSC, Mss 841505, 857518. For student exercise books with surveying problems from the 1860-70s. see DCLSC, DA-31. For earlier examples, see Sanborn C. Brown and Leonard M. Rieser, *Natural Philosophy at Dartmouth: From Surveyor's Chains to the Pressure of Light* (Hanover, New Hampshire: University Press of New England, 1974) 6-11.

168 Lord, ed., *Dartmouth College* 293-96, 423-29, 477-81; William Phelps Kimball, *The First Hundred Years of the Thayer School of Engineering at Dartmouth College* (Hanover, New Hampshire: University Press of New England, 1971); *Catalogue of the Officers and Students of Dartmouth College for the Academical Year 1873-4* (Hanover, New Hampshire: no publ., 1873) 68.

169 Constance E. Putnam, *The Science We Have Loved and Taught: Dartmouth Medical School's First Two Centuries* (Hanover, New Hampshire: University Press of New England, 2004) 16.

170 Trustee Minutes, August 1805, DCLSC, DA-1, vol. 1: 286; Oliver S. Hayward and Constance E. Putnam, *Improve, Perfect & Perpetuate: Dr. Nathan*

Smith and Early American Medical Education (Hanover, New Hampshire: University Press of New England, 1998) 77. See the student essays (probably delivered as commencement orations) on ignition of bodies in 1798 and 1799, on light in 1800, on the history of chemistry in 1810 and on relations between natural philosophy and chemistry in 1811, DCLSC, Commencement Parts Records, DA-43(1), and Student Essays, DA-101. In 1800, the College awarded an honorary degree to William Wilburforce, the well-known British politician, slavery abolitionist and philanthropist, and apparently mentioned its want of a "chemical apparatus." Wilburforce replied that the current "distress of the times" left him unable to meet the "demands of my own country," to say nothing of Dartmouth's. See Wilberforce, letter to Samuel Peters, 5 August 1801, DCLSC, Ms 801455.

171 Quoted in Brown and Rieser, *Natural Philosophy at Dartmouth* 27.

172 Cyrus Perkins, inventory, 28 September 1815, DCLSC, Ms 815528.4; Members of the Junior Class, letter to President Bennet Tyler, 6 October 1825, DCLS, Ms 825556; Foster, "Webster and Choate," 512. The trustees clearly distinguished the chemical from the philosophical apparatus, placing the former under the care of the medical professor for chemistry and the latter under the college professor of mathematics and natural philosophy. See Trustee Minutes, November 1814, DCLSC, DA-1, vol. 2: 14.

173 Stanley M. Guralnick, *Science and the Ante-Bellum American College* (Philadelphia: American Philosophical Society, 1975) 96-99. Harvard's first professor of chemistry was not named until 1783, a year after its medical school had been founded. See I. Bernard Cohen, "The Beginning of Chemical Instruction in America: A Brief Account of the Teaching of Chemistry at Harvard Prior to 1800," *Chymia* 3 (1950): 17-44.

174 Trustee Minutes, September 1815, August 1827, DCLSC, DA-1, vol. 2: 40, 236. In 1838, Dartmouth received its first major donation to support the sciences, $10,000 from Frederick Hall, Dartmouth 1803 and former professor of chemistry and natural philosophy at Trinity College, with one half providing for the purchase of Hall's mineralogical cabinet and the other half endowing a professorship in mineralogy and geology. See Lord, ed., *Dartmouth College* 258.

175 Oliver P. Hubbard, letter to Benjamin Silliman, 26 April 1836, DCLSC, Ms 836273. Joseph M. Wightman, *A Catalogue of Philosophical, Astronomical, Chemical and Electrical Apparatus* (Boston: Samuel N. Dickinson & Co., 1846) 10, includes Hubbard among the testimonials praising the firm's products. For Hubbard's purchasing patterns, see DCLSC, Ms 002050, and Treasurer's Papers, DCLSC, DA-2(4), folders 34-46.

176 Reuben Dimond Mussey, lecture notes, 1836; Cyrus Mentor Fisk, lecture notes, 1845; George Sullivan Gove, lecture notes, 1857-58, DCLSC, Vault 4.

177 Oliver P. Hubbard, letter to trustees, 14 August 1849, DCLSC, Ms 849464.

178 Trustee Minutes, August 1849, DCLSC, DA-1, vol. 3: 103; Oliver P. Hubbard, purchases for 1859-60, DCLSC, Ms 860424; *Catalogue of the Officers and Students of Dartmouth College for the Academical Year 1850-51* (Hanover, New Hampshire: Dartmouth Press, 1850) xxvii; Lord, ed., *Dartmouth College* 289; Putnam, *Dartmouth Medical School* 53-54.

179 *The Dartmouth* 3 (1869) 277-78; 5 (1871) 275-76; *Catalogue of the Officers and Students of Dartmouth College for the Academical Year 1871-72* (Boston: Rand, Avery & Co., 1871) 44; Lord, ed., *Dartmouth College* 347-59.

180 Edwin J. Bartlett, letter to President William J. Tucker, 23 February 1893, DCLSC, Tucker Papers, DP-9(47); *Dartmouth Bi-Monthly Magazine* 1 (1905): 10-11; Edwin J. Bartlett, "The Chemical Laboratory," *Dartmouth Alumni Magazine* November 1913: 6-9; *Catalogue of Dartmouth College, 1921-22* (Hanover: Printed for the College, 1921) 72; "Chemistry Materials Cost Total of $5500 Annually," *The Dartmouth* 23 April 1928: 4; Richardson, *History of Dartmouth College* 773-74.

181 *American Journal of Science* 6 (1823) 163-64, 330-31; 8 (1824) 275-76; James Freeman Dana, *An Epitome of Chymical Philosophy, Being an Extended Syllabus of the Lectures on That Subject, Delivered at Dartmouth College, and Intended as a Textbook for Students* (Concord, New Hampshire: Isaac Hill, 1825). For the suggestion that Dana owned his apparatus, see Reuben D. Mussey, letter to Benjamin Hale, 15 June 1827, DCLSC, Ms 827365. Dana's electromagnetic investigations had been prompted by the work of University of Pennsylvania chemist, Robert Hare, described in "Correspondence between Robert Hare...and the Editor on the Subject of Dr. Hare's Calorimotor and Deflagrator," *American Journal of Science* 5 (1822): 94-112. His 1827 public lectures on electromagnetism in New York attracted considerable attention. See James F. Dana, letter to Parker Cleaveland, 13 January 1826, DCLSC, Ms 826113; "Memoir of James Freeman Dana, M.D.," *Collections of the New Hampshire Historical Society* 2 (1827): 290-300; Samuel Irenaeus Prime, *The Life of Samuel F.B. Morse* (New York: D. Appleton and Company, 1875) 161-69, 267-69.

182 In 1842, Hubbard and his former mentor (and father-in-law) Silliman performed a chemical assay of a sample of coal. Although their article does not indicate where the analysis was performed, one might assume they used Yale's well-equipped laboratories. See Benjamin Silliman and O. P. Hubbard, "Chemical Examination of Bituminus Coal from the Pits of the Mid Lothian Coal Mining Company," *Amerian Journal of Science* 42 (1842): 369-74.

183 Bartlett co-authored two minor articles with undergraduate students (1895, 1897); Richardson co-authored two articles with other Dartmouth chemistry instructors (1913, 1923) and published another after a year of graduate studies at Cornell (1917). Neither Bartlett, Dartmouth 1872, nor Richardson, Dartmouth 1900, earned Ph.D. degrees. Scarlett, who did take a Ph.D. at Columbia, published that research (1917) and one other article with Berkeley chemists, after spending a sabbatical year there (1927).

184 Davis Baird, *Thing Knowledge: A Philosophy of Scientific Instruments* (Berkeley: University of California Press, 2004) 89-112.

185 Leon Burr Richardson and Andrew J. Scarlett, Jr., *A Laboratory Manual of General Chemistry* (New York: Henry Holt and Company, 1929) 133-34.

186 Lord, ed., *Dartmouth College* 1-174; Steven J. Novak, "The College in the Dartmouth College Case: A Reinterpretation," *New England Quarterly* 47 (1974): 550-63.

187 In a near riot, the college students managed to preserve the two fraternity libraries for their side. See Lord, ed., *Dartmouth College* 120-36.

188 In 1813-14, the College's entire expenditures totaled only $20,000; the president's annual salary was $850. An expenditure of $1,000 for apparatus thus represented a sizeable sum. Trustee Minutes, April 1811, August 1813, DCLSC, DA-1, vol. 1: 355, 357; vol. 2: 7; Treasurer's Papers, DCLSCS, DA-2(1), folders 56 and 73; Chase, *Dartmouth College* 629; Sara Schechner, "John Prince and Early American Scientific Instrument Making," *Publications of the Colonial Society of Massachusetts* 59 (1982): 431-503.

189 Miscellaneous receipts, Treasurer's Papers, DCLSC, DA-2(1), folders 22-25; Trustee Minutes, April and August 1819, DCLSC, DA-1, vol. 2: 115-16, 135; *Catalogue of the Officers and Students of Dartmouth College* (Boston: Phelps and Farnham, 1824) 18; Lord, ed., *Dartmouth College* 145, 176-78, 209.

190 Brown and Rieser, *Natural Philosophy at Dartmouth* 45-56. The nation's first professorship for "natural philosophy and astronomy" was created in 1836 for Denison Olmsted at Yale. See Deborah Jean Warner, "Astronomy in Antebellum America," *The Sciences in the American Context: New Perspectives,* ed. Nathan Reingold (Washington, D.C.: Smithsonian Institution Press, 1979) 63.

191 Ira Young, invoice, 20 August 1834, Treasurer's Papers, DCLSC, DA-2(4):11; Ira Young, letter to Ebenezer Adams, Jr., 11 May 1836, Ebenezer Adams Papers, DCLSC, ML-420(1):9. Of these early items acquired by Young, only a fragment of the alt-azimuth theodolite remains in the King Collection, signed by E. & G.W. Blunt of New York. R.G.W. Anderson, J. Burnett and B. Gee, *Handlist of Scientific Instrument-Makers' Trade Catalogues, 1600-1914* (Edinburgh: National Museums of Scotland, 1990) 97-98.

192 *Catalogue of the Officers and Students of Dartmouth College* (Newport, New Hampshire: Simon Brown, 1834) 20. Ira Young, letter to trustees, 22 July 1837, DCLSC, Ms 837422; student testimonials, DCLSC, Mss 898900 and 849212.

193 By 1841, Dartmouth had become one of the largest colleges in the East, graduating 76 students, in comparison to 78 at Yale, 60 at Princeton, and 48 at Harvard, according to Lord, ed., *Dartmouth College* 258.

194 Trustee Minutes, July 1845, DCLSC, DA-1, vol. 3: 47; Lord, ed., *Dartmouth College* 265-66. Samuel Appleton (1766-1853), born into a farming family in Ipswich, New Hampshire, moved to Boston in 1794 and became wealthy by importing British goods and buying into Lowell's cotton manufacture. In his later years, he devoted himself solely to philanthropy, annually giving away tens of thousands of dollars. "Throughout New England his name will be permanently connected with the charitable, educational and religious institutions which received aid from his ready and large-hearted munificence," wrote a eulogist in "Mercantile Biography: The Late Samuel Appleton, Esq.," *Western Journal and Civilian* 11 (1854): 431. Appleton always presented himself as a simple, self-made man and undoubtedly would have respected Young's humble origins. Dartmouth awarded Appleton an honorary degree in 1849.

195 Young had visited most of these institutions during his 1834 tour. He knew Western Reserve College, in Hudson, Ohio, though his relationships with Elias Loomis, professor of mathematics and natural philosophy at that college from 1837-44, and with Clement Long, a member of Young's Dartmouth class of 1828 and professor of moral philosophy and Christian theology at Western Reserve since 1834. See Ira Young, letter to Elias Loomis, 11 April 1839, Loomis Papers, Beinecke Rare Book and Manuscript Library, Yale University, Z117.00113 v. 12.

196 Faculty Committee, letter to the trustees, 27 July 1846, DCLSC, Ms 846427.1; Ira Young, letter to the trustees, 28 July 1846, DCLSC, Shattuck Observatory Papers, DA-9(1):12; trustee committee report, undated, DCLSC, Ms 846429; Trustee Minutes, July 1846, DCLSC, DA-1, vol. 3: 64-65; Lord, ed., *Dartmouth College* 264; Richardson, *History of Dartmouth College* 402. For contemporary and surprisingly similar visions for laboratory science at German universities, see Richard L.

Kremer, "Building Institutes for Physiology in Prussia, 1836-1846: Contexts, Interests and Rhetoric," *The Laboratory Revolution in Medicine,* eds. Andrew Cunningham and Perry Williams (Cambridge: Cambridge University Press, 1992) 72-109.

197 George Shattuck, letter to Ebenezer Adams, 6 July 1821, Adams Papers, DCLSC, ML-420(2):15; Young, letters to George Shattuck, 27 August 1852 and 23 September 1852, Shattuck Papers, Massachusetts Historical Society, Ms N-909, vol. 22; Harvard invoices, September 1852, Treasurer's Papers, DCLSC, DA-173(3).

198 Young, letters to George Shattuck, 2 and 5 October 1852, copied in Trustee Minutes, December 1852, DCLSC, DA-1, vol. 3: 155-70. Cf. Brown and Rieser, *Natural Philosophy at Dartmouth* 49-54; Deborah Jean Warner, "Projection Apparatus for Science in Antebellum America," *Rittenhouse* 6 (1992): 87-94. For the Amherst collection, which Young knew well, see Thomas B. Greenslade, Jr., "Collection Profile: Visits to Apparatus Collections III–Amherst College," *Rittenhouse* 15 (2001): 39-46.

199 Lord Dartmouth, letter to President Lord, 21 May 1853, DCLSC, Ms 853321. See David Pantalony, "The Purchasing Trip of Ira and Charles Young in 1853," *Bulletin of the Scientific Instrument Society* 76 (2003): 23-27; Chandos Michael Brown, *Benjamin Silliman: A Life in the Young Republic* (Princeton: Princeton University Press, 1989).

200 Silliman, letter of introduction, 15 April 1853, Charles A. Young Papers, DCLSC, ML-49(7):113; Charles A. Young, "Diary of a Trip to Europe," 1853, autograph ms., Charles A. Young Papers, DCLSC, ML-49(1):14, 15.

201 Charles Young, letter to George C. Shattuck, 2 June 1853; Ira Young, letter to George C. Shattuck, 16 August 1853, Shattuck Papers, Massachusetts Historical Society, Ms N-909(22); Angelo Secchi, letter to Ira Young, 10 June 1853, Charles Young Papers, ML-49(7):109; Ira Young, undated report for trustees [July 1854], DCLSC, Ms 852940; Young, "Diary of a Trip to Europe." See Deborah Jean Warner, "French Instruments in the United States," *Rittenhouse* 8 (1993): 1-32.

202 Meteorological registers, Shattuck Observatory Papers, DCLSC, DA-9(2):4 and 6; Ira Young, "Record of Observations Chiefly with the Transit Instrument at the Dartmouth College Observatory," 1850-51, notebook, DLCSC, Ms 003059; John K. Lord, *A Discourse Commemorative of Ira Young* (Hanover, New Hampshire: Dartmouth Press, 1859). In 1830, Young did author several sections of a popular mathematics school book. See Daniel Adams, *Adam's New Arithmetic...Designed for the Use of Schools and Academies in the United States* (Keene, New Hampshire: J. & J.W. Prentiss, 1830).

203 Trustee Minutes, January 1862, November 1863, DCLSC, DA-1, vol. 3: 306-7, 344; Treasurer's report, 1861-62, Treasurer's Papers, DCLSC, DA-2(5):10; Lord, ed., *Dartmouth College* 320.

204 Henry Fairbanks, "Appleton Fund Inventory," 1862, autograph ms., Treasurer's Papers, DCLSC, DA-2(81):12, 24. By comparison, an 1867 inventory of the Amherst collection lists 612 items. See Greenslade, "Amherst College." Somatology refers to the science of bodies, including cohesion, adhesion, capillarity, osmotic pressure.

205 Nearly a dozen of the electro-magnetic instruments may correspond to apparatus found in Daniel Davis, Jr., *Manual of Magnetism,* 3d ed. (Boston: Daniel Davis, Jr., 1851). Davis is known not to have signed his early instruments. Had Young purchased from Davis or copied his designs?

206 President Asa D. Smith, letter to Charles Young, 17 July 1865, Young Papers, DCLSC, ML-49(2):80; Charles Young, letter to Asa D. Smith, 8 September 1865, DCLSC, Ms 865508; Faculty salaries, 1868-69, DCLSC, Ms 868900.3; *The Dartmouth* 3 (1869) 399; 4 (1870) 153; 5 (1871) 324-25; 6 (1871) 117; 6 (1872) 40; *Catalogue of the Officers and Students of Dartmouth College* (Hanover, New Hampshire: Dartmouth College, 1869) 46; James Patterson, "Address on the Relations of the College to Science and the Arts," *Centennial Celebration at Dartmouth College, July 21, 1869* (Hanover, New Hampshire: J.B. Parker, 1870) 61-70. In 1871 after Young had turned down an invitation from the British astrophysicist, Norman Lockyer, to join an international eclipse expedition to India, a writer in *The Dartmouth* 5 (1871) 370, worried: "Nor are we so sure that [Young's] loss is all our gain, for highly as we value his instruction we feel that such an appreciation of merit from such a source is an honor to Dartmouth as well as to Dr. Young himself. In July last Prof. Young was elected an associate Fellow of the American Academy (at Boston), and also received the honorary degree of Ph.D. from Hamilton College."

207 Trustee Minutes, July 1866, DA-1, vol. 3: 401; Charles Young, Diaries, 1866, 1869, Young Papers, DCLSC, ML-49(1):8; receipts and invoices, Appleton Fund Papers, DCLSC, DA-173; Treasurer's Papers, DA-2(5):16-19; Charles A. Young, "Induction Coil of Unusual Size," *Journal of the Franklin Institute* 60 (1870): 7.

208 Receipts, 1870-72, Appleton Fund Papers, DCLSC, DA-173(2); *The Dartmouth* 6 (1872) 86-87; Charles A. Young, "Spectroscopic Observations," 1870-72, ms. notebook, DCLSC, ML-49(1):19. The students did not mention Young's largest European purchase, a compound microscope and accessories by Henry Crouch of London that had cost nearly $1,000 and would arrive later in Hanover due to delays in its manufacture. Young's 1870-71 purchases include an electric egg and induction coil by Apps and carbon bisulfide prisms and direct-vision spectroscopes by Browning (see entries).

209 *Catalogue of the Officers and Students of Dartmouth College* (Hanover, New Hampshire: Dartmouth College, 1872) 44. Dartmouth's students petitioned the trustees, urging that Young be retained. The trustees offered him $5,000 for additional observatory equipment and a new chair in astronomy, to be endowed for $45,000. See Trustee Minutes, February 1877, DCLSC, DA-1, vol. 4: 181-84; *The Dartmouth* 2 (1877) 271-72, 287. Already in 1875, Young had received offers from other institutions. See Asa Smith, letter to Thaddeus Fairbanks, 3 August 1874, DCLSC, Ms 875453.

210 *Exercises at the Inauguration of Samuel Colcord Bartlett, D.D., as President of Dartmouth College* (Concord, New Hampshire: Republican Press Association, 1877); Lord, ed., *Dartmouth College* 413-37; Brown and Rieser, *Natural Philosophy at Dartmouth* 85-100. In 1893, the title of Emerson's chair was changed from natural philosophy to physics; in 1898, he became the first dean of the faculty and taught no more science.

211 Ganot's textbook appeared in 29 French editions from 1851-1928; 25 Italian editions through 1896; 18 Spanish editions through 1923; 18 London editions through 1910; 18 New York editions through 1917; one Turkish edition in 1876; one Bulgarian edition in 1869; and significantly only one German edition (1858). Emerson's personal copy of the 1872 edition (DCLSC, Rumford QC21 .G2 1872) is bound with interleaved blank sheets. He divided the text into 83 sections for lectures, updated it by copying paragraphs and several dozen engravings from the 1875 6th edition, and corrected typographical and other errors. See Benjamin Silliman, Jr., *Principles of Physics, or Natural Philosophy Designed for the Use of Colleges and Schools*, 2d rev. ed. (Philadelphia: Peck & Bliss, 1861); Adolphe Ganot, *Elementary Treatise on Physics, Experimental and Applied for the Use of Colleges and Schools*, trans. E. Atkinson, 5th ed. (New York: William Wood and Co., 1872); *Dartmouth Catalogue 1872*, 42; Charles Franklin Emerson, "The Seven Founders: An Address Commemorating the Fiftieth Anniversary of the Dartmouth Scientific Association," 1920, ts., DCLSC, DC Hist Q11 .D3.E42, 3 (Emerson); Josep Simon, "Adolphe Ganot (1804-1887) and His Textbook of Physics," M.Sc., University of Oxford, 2004.

212 *Catalogue of Dartmouth College and the Associated Institutions* (Hanover, New Hampshire: Dartmouth College, 1882) 22-23. The required textbooks include Alfred P. Gage, *A Textbook on the Elements of Physics for High Schools and Academics* (Boston: Ginn, Heath & Co., 1882); Edward C. Pickering, *Elements of Physical Manipulation*, 2 vols. (New York: Hurd & Houghton, 1873-76); James Clerk Maxwell, *Matter and Motion* (New York: Pott, Young, 1876); William H. Stone, *Elementary Lessons on Sound* (London: Macmillan & Co., 1879); Balfour Stewart, *An Elementary Treatise on Heat* (Oxford: Clarendon Press, 1866); Eugene Lommel, *The Nature of Light, with a General Account of Physical Optics* (London: H.S. King & Co., 1875); Fleeming Jenkin, *Electricity and Magnetism* (London: Longmans, Green, and Co., 1873). Professor Emerson had assisted Gage, Dartmouth 1859, in the preparation of his textbook. See Steven Turner, "The Reluctant Instrument Maker: A.P. Gage and the Introduction of the Student Laboratory," *Rittenhouse* 18.2 (2004): 49.

213 *The Dartmouth* 5 (1883) 10. Optional exercises in "practical chemistry" had been offered occasionally since 1865-66, and in "practical astronomy" since 1866-67. From 1879-80 until the elective reforms, optional exercises in physics, chemistry and astronomy were offered to seniors.

214 See Stanley M. Guralnick, "The American Scientist in Higher Education, 1820-1910," in Reingold, ed., *New Perspectives*, 118-20.

215 Pickering, *Physical Manipulation* 1: vi.

216 Turner, "A.P. Gage," 41-61.

217 Student grade book, 1878-85, DCLSC, DA-80.1(11); *Catalogue of Dartmouth College and the Associated Institutions* (Hanover, New Hampshire: Dartmouth College, 1885) 23, 30; *Catalogue of Dartmouth College* (Hanover, New Hampshire: Dartmouth College, 1893) 63-64; Paul P. Bénézet, "The History of Reed Hall," 1932, ts., DCLSC, DC History LD1440 .F4 .B4, 8.

218 Trustee Minutes, June 1879, DCLSC, DA-1, vol. 4: 241; Apparatus of the Appleton Professorship, undated inventory, DCLSC, Ms 000737. In 1886, Emerson replaced Pickering with a laboratory manual written by demonstrators at the Cavendish Laboratory in Cambridge, Richard T. Glazebrook and W.N. Shaw, *Practical Physics* (New York: D. Appleton, 1885). See Thomas B. Greenslade, Jr., "Home Built Apparatus for Natural Philosophy," *Rittenhouse* 7 (1993): 56-59.

219 William Jewett Tucker, *The Historic College, its Present Place in the Educational System* (Hanover, New Hampshire: Dartmouth College, 1894) 21, 34. Cf. Paul Forman, John L. Heilbron and Spencer Weart, "Physics Circa 1900: Personnel, Funding, and Productivity of the Academic Establishments," *Historical Studies in the Physical Sciences* 5 (1975): 1-185; Stanley Goldberg and Roger H. Stuewer, eds., *The Michelson Era in American Science, 1870-1930* (New York: American Institute of Physics, 1988); David Cahan and M. Eugene Rudd, *Science at the American Frontier: A Biography of Dewitt Bristol Brace* (Lincoln: University of Nebraska Press, 2000).

220 Edwin B. Frost, "Experiments on the X-Rays," *Science* N.S. 3 (1896): 235-36; Albert Cushing Crehore, *Autobiography* (Gates Mills, Ohio: William G. Berner, 1944). Crehore was the first Dartmouth physics instructor to have earned a Ph.D. (Cornell);

Frost was the first Dartmouth professor to publish in *Science*.

221 "Wilder Physical Laboratory," *The Dartmouth* 19 (1898) 308-10; Ernest Fox Nichols, "The Wilder Physical Laboratory of Dartmouth College," *Physical Review* 12 (1901): 366-71; Andrew D. Field, "The Construction of a Physical Laboratory and the Evolution of a Physics Department, Dartmouth College, 1880-1910," 1991, Wilder Laboratory Vertical File, DCLSC.

222 Ernest Fox Nichols, letter to President William J. Tucker, 30 June 1898, Tucker Papers, DCLSC, DP-9(6):41; E.F. Nichols and Gordon Ferrie Hull, "Pressure Due to Radiation," *Physical Review* 13 (1901): 307-20; Ernest Fox Nichols and Gordon Ferrie Hull, "The Pressure Due to Radiation," *Astrophysical Journal* 17 (1903): 315-51; Field, "Physical Laboratory," 4-5. For apparatus used by Nichols and Hull, see entries for the ruling engine, Wheatstone bridge, and astatic mirror galvanometer.

223 Ernest Fox Nichols, "Instrument Purchases," 1898-1908, ms. notebook, King Collection Archive.

224 Trustee Minutes, May 1903, DCLSC, DA-1, vol. 5: 323-24 (italics added). Hull later recalled that he and Nichols had done the pressure of light experiments late at night, after having "spent full days on duties connected with teaching." Gordon Ferrie Hull, "Reminiscences of a Scientific Comradeship," *American Physics Teacher* 4 (1936): 61-65; Field, "Physical Laboratory," 5.

225 Gordon F. Hull, letter to John King Lord, 18 December 1912, Physics Department Vertical File, DCLSC.

226 Deborah Jean Warner, "What Is a Scientific Instrument, When Did It Become One, and Why?," *British Journal for the History of Science* 23 (1990): 83-93; Willem Hackmann, "Scientific Instruments: Models of Brass and Aids to Discovery," *The Uses of Experiment*, eds. David Gooding, Trevor Pinch and Simon Schaffer (Cambridge: Cambridge University Press, 1990) 31-66; Patrick Carroll-Burke, "Tools, Instruments and Engines," *Social Studies of Science* 21 (2001): 593-625; Baird, *Thing Knowledge*.

227 Anita McConnell, "From Craft Workshop to Big Business: The London Scientific Trade's Response to Increasing Demand, 1750-1820," *London Journal* 19 (1994): 36-53.

228 *Catalogue of Mathematical and Optical Instruments, Philosophical and Chemical Apparatus, Constructed and Sold by James Green* (Baltimore: no publ., 1844) 15; Benjamin Pike, *Illustrated Descriptive Catalogue of Optical, Mathematical and Philosophical Instruments,* 2 vols. (New York: By the author, 1848) 2: 272-77.

229 Deborah Jean Warner and Robert B. Ariail, *Alvan Clark & Sons: Artists in Optics*, 2d ed. (Richmond, Virginia: Willmann-Bell, Inc., 1995) 208.

230 David Pye, *The Nature and Aesthetics of Design*, rev. ed. (New York: Van Vostrand Reinhold, 1978).

Sources of Illustrations

DCL = Dartmouth College Library; TDC = Trustees of Dartmouth College. Images of items in the King Collection are not part of this list and are reproduced courtesy of the TDC.

Frontis. DCLSC, Photographic Records, 10-47-192. Courtesy of DCL.

P.5 Edwin Thacher, *Directions for Using Thacher's Calculating Instrument* (New York: Keuffel & Esser Co., 1907) title page. Courtesy of the TDC.

P.6 Thacher, *Directions* 72. Courtesy of the TDC.

P.8 Central Scientific Company, *Scientific Instruments, Laboratory Apparatus and Supplies for Physics, Chemistry, Biological Sciences, and Industrial Testing, Catalogue J136* (Chicago: Neely Printing Company, 1936) 87. King Collection Archive. Courtesy of the TDC.

P.16 Undated ephemeral advertisement, Laboratory for Science, King Collection Archive. Courtesy of the TDC.

P.18 George Adams, *Lectures on Natural and Experimental Philosophy*, 4 vols. (Philadelphia: Whitehall, 1807) 3: Pl. 5. Courtesy of DCL.

P.20 Clausing vertical mill, Model 8520. Courtesy of Francis J. Manasek.

P. 37 William Bond & Son, "Daybook," 1833-35, ms. notebook, Records of William Bond & Son and the Bond Family, 1724-1931, Scientific Instrument Collection, Harvard University, box 4, folder 2. Courtesy of the Collection of Historical Scientific Instruments at Harvard University.

P.40 Fairchild Type A-10 sextant instructions, April 1944. King Collection Archive. Courtesy of the TDC.

P.41 Richard H. Goddard, undated drawing [about 1947], King Collection Archive. Courtesy of the TDC.

P.47 Eastern Science Supply Company, *ESSCO Rapid Refernce Pictorial Bulletin, RR1* (Boston: Eastern Science Supply Company, 1932) 4. King Collection Archive. Courtesy of the TDC.

P.48 H.W. Geromanos, letter (with photograph) to Richard H. Goddard, 27 October 1937, King Collection Archive. Courtesy of the TDC.

P.49 S.L. Boothroyd, letter to Norman E. Gilbert, 3 May 1940. King Collection Archive. Courtesy of the TDC.

P.50 DCLSC, Photographic Records, 10-47-182. Courtesy of DCL.

P.53 Kullmer Advertisement, *The Public* 14 (1911): 1088. Clipping from the Charles J. Kullmer Papers, Archives and Records Management, Syracuse University. Courtesy of Syracuse University. All rights reserved.

P.54 William Simms, *The Achromatic Telescope and its Various Mountings* (London: Troughton & Simms, 1852) 26. Courtesy of Francis J. Manasek.

P.56 DCLSC, Photo File, Astronomy. Courtesy of DCL. Charles A. Young, *The Sun* (New York: D. Appleton and Company, 1881) 51. Courtesy of DCL.

P.57 DCLSC, Photo File, Observatory II. Courtesy of DCL.

P.58 Courtesy of Francis J. Manasek.

P.59 *Johnson's New Illustrated Family Atlas* (New York: Johnson & Ward, 1862). Courtesy of Francis J. Manasek.

P.60 DCLSC, Photo File, Observatory Apparatus, H.O. Bly photograph. Courtesy of DCL.

P.62 King Collection Archive, Telespectroscope file, H.O. Bly photograph. Courtesy of the TDC.

P.63 Wilhelm Struve, "An Account of the Arrival and Erection of Fraunhofer's Large Refracting Telescope at the Observatory of the Imperial University at Dorpat," *Memoirs of the Astronomical Society of London* 2 (1826): plate. Courtesy of DCL.

P.64 DCLSC, Photo File, Observatory Apparatus, H.O. Bly photograph. Courtesy of DCL.

P.65 King Collection Archive, Telespectroscope file, H.O. Bly photograph. Courtesy of the TDC.

P.68 DCLSC, Photo File, Observatory Apparatus, H.O. Bly photograph. Courtesy of DCL.

P.69 Henry J. Green, letter to Richard H. Goddard, 15 June 1934, King Collection Archive. Courtesy of the TDC.

P.70 DCLSC, Photo File, Observatory I. Courtesy of DCL.

P.73 Clark receipt, 2 March 1872, Shattuck Observatory Papers, DCLSC, DA-9(1):9. Courtesy of DCL.

P.74 *Bulletin of the U.S. Coast and Geodetic Survey* 1 (1888): Appendix 15, Pl. 50. Courtesy of Francis J. Manasek.

P.77 Undated ephemeral advertisement [about 1950], King Collection Archive. Courtesy of the TDC. SurpluShed, *Surplus Optics & Electronics Catalogue* (Blandon, Pennsylvania: no publ., 2004) 95. Courtesy of SurpluShed.

P.80 Young, *The Sun* 86. Courtesy of DCL.

P.82 Charles A. Young, "Spectroscopic Notes," *Nature* 3 (1870): 110. Courtesy of DCL.

P.84 Young, *The Sun* 190. Courtesy of DCL.

P.86 "Reports of Observations upon the Total Solar Eclipse of December 22, 1870," *U.S. Coast and Geodetic Survey Report for 1870, Appendix No. 16,* ed. Benjamin Peirce (Washington, D.C.: Government Printing Office, 1870) unnumbered Pl. Courtesy of DCL.

P.87-88 Heinrich Schellen, *Spectrum Analysis in its*

Application to Terrestrial Substances and the Physical Constitution of the Heavenly Bodies [1871], trans. Jane Lassell and Caroline Lassell, ed. William Huggins (London: Longmans, Green & Co., 1872) 479, 119, 90. Courtesy of DCL.

P.92 Eastern Science Supply Company, *Rapid Reference* title page. King Collection Archive. Courtesy of the TDC.

P.93 Tyrer & Co. Invoice, 20 April 1906, King Collection Archive. Courtesy of the TDC.

P.94 Sir William Ramsay's Instructions (London: Thomas Tyrer & Co. [c. 1906]), King Collection Archive. Courtesy of the TDC.

P.96 Gordon Ferrie Hull, undated ms., King Collection Archive. Courtesy of the TDC.

P.101 R. & J. Beck, *An Illustrated Catalogue of Microscopes and Other Scientific Instruments* (Philadelphia: no publ., 1879) front cover. Courtesy of the Collection of Historical Scientific Instruments at Harvard University.

P.109 William Henry Hall, *New Royal Encyclopedia*, 3 vols. (London: C. Cooke, 1789-91) Pl. 1. Courtesy of Francis J. Manasek.

P.117 Richard H. Goddard, "New Telescope," *Dartmouth Alumni Magazine* April 1939: 28. Courtesy of TDC.

P.119 L.E. Knott Apparatus Company, *Catalogue of Scientific Instruments, Catalogue 17* (Boston: L.E. Knott Apparatus Company, 1912) 322. Courtesy of the Smithsonian Institution Libraries.

P.121 W. & L.E. Gurley Invoice, 30 October 1908, King Collection Archive; W. & L.E. Gurley, *Physical and Scientific Instruments* (Troy, New York: W. & L.E. Gurley, 1910) 144. King Collection Archive. Courtesy of the TDC.

P.123 Hall, *Encyclopedia* Pl. 2. Courtesy of Francis J. Manasek.

P.126 Adolphe Ganot, *Elementary Treatise on Physics, Experimental and Applied for the Use of Colleges and Schools*, trans. E. Atkinson, 5th ed. (New York: William Wood and Co., 1872) 143. Courtesy of the DCL.

P.129 DCLSC, Photo File, Scientific Apparatus, H.O. Bly photographs. Courtesy of DCL.

P.130 Paul P. Bénézet, "The History of Reed Hall," 1932, ts., DCLSC, DC History LD1440 .F4 .B4, unpag. appendix. DCLSC, Photo File, Wilder Laboratory. Courtesy of DCL.

P.131 Ernest Fox Nichols, "The Wilder Physical Laboratory of Dartmouth College," *Physical Review* 12 (1901): 368-69. DCLSC, Photo File, Wilder Laboratory. Courtesy of DCL.

P.133 Central Scientific Company, *Catalogue J136* 1575. King Collection Archive. Courtesy of the TDC.

P.134 DCLSC, Photographic Records, 6-62-77. Courtesy of DCL.

P.135 DCLSC, Photographic Records, 6-55-252. Courtesy of DCL.

P.138 Rudolph Koenig, *Quelques expériences d'acoustique* (Paris: A. Lahure, 1882) 57, 52. Courtesy of DCL.

P.139 DCLSC, Photo File, Scientific Apparatus, H.O. Bly photograph. Courtesy of DCL.

P.142 Adolphe Ganot, *Traité élémentaire de physique expérimentale et appliquée et de météorologie,* 6th ed. (Paris: L'auteur-éditeur, 1856) 202. Courtesy of DCL.

P.143 Ernst Florens Friedrich Chladni, *Die Akustik* (Leipzig: Breitkopf und Härtel, 1802) title page. Courtesy of DCL.

P.145 Max Kohl A.G., *Physical Apparatus, Price List No. 50* (Chemnitz: 1911) v. King Collection Archive. Courtesy of the TDC.

P.146 Max Kohl A.G., *Price List No. 50* 416. King Collection Archive. Courtesy of the TDC.

P.148-49 George Adams, *An Essay on Electricity*, 3d ed. (London: R. Hindmarsh, 1787) Pl. 1, frontispiece. Courtesy of DCL.

P.154 Adolphe Ganot, *Elementary Treatise on Physics, Experimental and Applied for the Use of Colleges and Schools*, trans. E. Atkinson, 15th ed. (New York: William Wood and Company, 1901) 821. Courtesy of DCL.

P.157 Charles G. Page, "Magneto-Electric and Electro-Magnetic Apparatus and Experiments," *American Journal of Science* 35 (1839): 258. Courtesy of DCL.

P.158 Charles G. Page, "On the Use of the Dynamic Multiplier," *American Journal of Science* 32 (1837): 356. Courtesy of DCL.

P.159 Daniel Davis, Jr., *Manual of Magnetism* (Boston: Daniel Davis, Jr., 1842) 190. Courtesy of Niels Bohr Library, Center for History of Physics, American Institute of Physics.

P.162 Elroy McKendree Avery, *School Physics* (New York: American Book Company, 1895) 506. Courtesy of Francis J. Manasek.

P.163 DCLSC, Photo File, Scientific Apparatus. Courtesy of DCL.

P.164 Schellen, *Spectrum Analysis* 25. Courtesy of DCL.

P.167 Adolphe Ganot, *Elementary Treatise on Physics, Experimental and Applied for the Use of Colleges and Schools*, 13th ed. (London: Longmans, Green, and Co., 1890) 931. King Collection Archive. Courtesy of the TDC.

P.169-70 Allen & Hanburys, *Catalogue of Surgical Instruments and Appliances, Aseptic Hospital Furni-*

ture, Ward Requisites, Etc. (London: Allen & Hanburys, 1901) 831, 833. Courtesy of Francis J. Manasek.

P.171 DCLSC, Photo File, X-ray pictures. Courtesy of DCL.

P.173 DCLSC, Photographic Records, 10-47-188. Courtesy of DCL.

P.174 W. G. Pye & Co., *Catalogue of Scientific Apparatus* (Cambridge: A. P. Dixon, Printer, 1911) 64. King Collection Archive. Courtesy of the TDC.

P.176 Copper utensils, Eastern Anatolia, late Ottoman. Courtesy of Frank J. Manasek.

P.182 DCLSC, Photo File, Physics Laboratory. Courtesy of DCL.

P.187 Leeds & Northrup Company, *Lamp and Scale Reading Device, No. 2100, Direction Book 77-6-0-16* (Philadelphia: Lithograph, 1939) 4. King Collection Archive. Courtesy of the TDC.

P.188 Leeds & Northrup Company, *Electrical Measuring Instruments for Research, Teaching and Testing* (Philadelphia: no publ., 1939) 40. King Collection Archive. Courtesy of the TDC.

P.189 *Students' Potentiometer, Bulletin No. 765*, (Philadelphia: Leeds & Northrup Company, 1926) front cover. King Collection Archive. Courtesy of the TDC.

P.191 Cambosco Scientific Company, *Cambosco Order Book, 1950-51* (Boston: Cambosco, 1950) front cover. Courtesy of the Physics Museum, University of Vermont.

P.194 General Radio Company, *Handbook of High-Speed Photography* (West Concord, Massachusetts: General Radio Company, 1963) front cover. King Collection Archive. Courtesy of the TDC.

P.195 *General Radio Experimenter* September 1960: front cover. Courtesy of the General Radio Historical Society.

P.197 DCLSC, Photo File, Physics Laboratory. Courtesy of DCL.

P.200 DCLSC, Photographic Records, 7-66-211. Courtesy of DCL.

P.211 DCLSC, Photo File, Chandler Scientific Department. Courtesy of DCL.

P.214 DCLSC, Photo File, Chemistry Laboratory (Old Lab), from H.O. Bly negative #37. Courtesy of DCL.

P.215 DCLSC, Photo File, Chemistry Laboratory (Old Lab). Courtesy of DCL.

P.217 Leon Burr Richardson and Andrew J. Scarlett, Jr., *A Laboratory Manual of General Chemistry* (New York: Henry Holt and Company, 1929) 133-34. Courtesy of the DCL.

P.222-23 DCLSC, Ms 846427.1. Courtesy of DCL.

P.225 DCLSC, Shattuck Observatory Papers, DA-9(1):1. Courtesy of DCL.

P.228 DCLSC, Charles A. Young Alumnus File, H.O. Bly photograph. Courtesy of DCL.

P.233 DCLSC, Gordon Hull Papers, ML-47(10):8. Courtesy of DCL.

P.238 Courtesy of Francis J. Manasek.

Index

Johnson, E. F. Company 199
Johnson, Richard C. 33
Johnson, Samuel 34
Johnston, W. & A.K. Ltd. 46
Jones, W. & S. 45, 54, 76
Jones, William 210
Journal of the Franklin Institute 227

K

Kagiya Yoko 42
Keiser & Schmidt 178
Kennard, John 24
Keuffel & Esser Company 5
King, Allen
        12, 33, 60, 81, 162, 199, 200, 204, 237
King, Charles Gedney 29
King's College 207
Kirchhoff, Gustav 79, 164
Kleist, Ewald 152
Knapp, Hermann Jakob 99
Knott, L.E. Apparatus Company
        7, 118, 123,
        185, 234
Knott, Louie E. 118
Koenig, Rudolph 137, 145
Koenig sound analyzer 137
Kohl, Max 145, 178
Kolbe, Bruno 119
Kullmer, Charles Julius 51

L

laboratory courses 230
Laboratory for Science 15
Ladd battery 154
Ladd, W. & Company 154
Lafayette Radio 75
lamp, carbon-filament 192
lamp, fluorescent 164
lamp, incandescent 192
Landolt, Edmond 99
Langill, Howard H. 171
LASER 199
Lassell, William 62
Lavoisier, Antoine 10
lead screw 97
Leeds & Northrup Company
        181, 186, 188, 234
Leitz 109
lens, achromatic 71
level 32
Leyden battery 151

Leyden jar 148, 152, 190, 195
light beam, amplifying movement 187
light, monochromatic 199
light, polarized 199
light, pressure of 97, 178, 233
lines, spectral 79
Littrow, Otto von 82
Locke, John Dexter 105
Lockyer, J. Norman 80, 94
London Polytechnic 161
Loomis, Elias 58
Lord, Nathan 226
Los Alamos 197
Lovely, John W. 116
Ludwig, Carl 113
lunar distance 35, 38
Lundin, Carl A.R. 64, 71
lute 10
lycopodium powder 143

M

Macalaster Wiggin Company 168
Machine Age 185
machine shop 50
machinists, signing work 20
magic lantern 108
magnet 15
magnetic field, earth's 177
magnetic meridian 177
magnetic observatory, Williams College 221
Maiman, Theodore Harold 199
making instruments 239
manganin 184
Mann, David W. 11
Mann, David W. Company 11
Mann Instrument Company 11
Mann short-interval timer 11, 237
Mannheim slide rule 5
manometer 124
manual, 1929 laboratory 217
Margetts, George 35
Marloye et Cie 140, 143, 224
Marquarts Lager chemischer Utensilien 9
Martin, Benjamin 208, 210
Mary Hitchcock Memorial Hospital 173
MASER 199
Maser Optics Inc. 198
Masonic emblem (on compass) 25
Mathieu, Roger 115
Maxwell, James Clerk 233
McCarthy, Eddie 171

267

Colophon

This book was manufactured in the United States of America.
The body of the text was set using the Adobe family of Caslon type.
The book is printed on acid-free paper and the gatherings are Smythe-sewn.
Alan Berolzheimer is the copy editor and proofreader.
The jacket design is by Carrie Fradkin of C Design, Lebanon, New Hampshire.
Layout and book design was done at Terra Nova Press in Norwich, Vermont.

MMV

Nordovicum